Science and Political Controversy

Recent Titles in the
CONTEMPORARY WORLD ISSUES
Series

Books in the **Contemporary World Issues** series address vital issues in today's society such as genetic engineering, pollution, and biodiversity. Written by professional writers, scholars, and nonacademic experts, these books are authoritative, clearly written, up-to-date, and objective. They provide a good starting point for research by high school and college students, scholars, and general readers as well as by legislators, businesspeople, activists, and others.

Each book, carefully organized and easy to use, contains an overview of the subject, a detailed chronology, biographical sketches, facts and data and/or documents and other primary source material, a directory of organizations and agencies, annotated lists of print and nonprint resources, and an index.

Readers of books in the **Contemporary World Issues** series will find the information they need in order to have a better understanding of the social, political, environmental, and economic issues facing the world today.

Science and Political Controversy

A REFERENCE HANDBOOK

David E. Newton

 ABC-CLIO

Santa Barbara, California • Denver, Colorado • Oxford, England

Copyright 2014 by ABC-CLIO, LLC

Library of Congress Cataloging-in-Publication Data

Newton, David E.
    Science and political controversy : a reference handbook / David E. Newton.
        pages cm. — (Contemporary world issues)
    Includes index.
    ISBN 978-1-61069-319-6 (hardback) —
ISBN 978-1-61069-320-2 (ebook)
1. Science—Political aspects.   2. Science and state—History.
3. Science—Political aspects—United States.   4. Science and state—United States.   I. Title.
    Q175.5.N469   2014
    338.9'26—dc23   2013048129

ISBN: 978-1-61069-319-6
EISBN: 978-1-61069-320-2

18 17 16 15 14   1 2 3 4 5

This book is also available on the World Wide Web as an eBook.
Visit www.abc-clio.com for details.

ABC-CLIO, LLC
130 Cremona Drive, P.O. Box 1911
Santa Barbara, California 93116-1911

This book is printed on acid-free paper ∞
Manufactured in the United States of America

During the first decade of the 21st-century, a number of scientists and laypersons expressed concerns as to what they perceived as the politicization of science by some federal agencies and federal bureaucrats. They pointed to reports that government officials had pressured scientists employed by the federal government to select or carry out certain research projects based on political factors rather than scientific criteria; had pressured scientists to revise or reinterpret their findings in ways that would be more supportive of political positions than scientifically justifiable; and had even rewritten some scientific reports to reflect governmental policies rather than representing accurate scientific information. As news of these events became known to the general public, a few officials accused of such wrongdoing resigned their positions, sometimes acknowledging that they believed that their primary responsibility was to the government they served, and not necessarily to the veracity of the scientific reports for which their agencies were responsible.

A review of this period of history reveals two important points. First, although many critics pointed to the involvement of one political party in these wrongdoings, it turns out that both political parties—and perhaps politicians in general—were sometimes motivated by political factors in making decisions on science-based issues. Second, the furor over the politicization of science in the 21st century appeared often to ignore the fact that such activities have a very long history, dating to at least the birth of modern science in the 16th century. Indeed, as early as the third century BCE, one of the world's most highly regarded philosophers was teaching that it is sometimes in the best interest of rulers to lie to their constituents because such lies (which he called *noble lies*) worked for the better good of the people as a whole. In any case, history is replete with examples of governments' manipulating the process of scientific research and the work of scientists for their own benefit. And examples also exist of the way in which scientists and their supporters reverse that process, using public resources for the betterment of science.

This book describes the process by which these two events take place, the politicization of science and the scientification (if such a term may be used) of politics. The book begins with a review of some examples of these processes throughout the pre–World War II history, including the use of demonstrably false biological theories during the 1930s in the Soviet Union for the development of a national plan for agriculture, a plan that ultimately proved disastrous to the nation, and the wholesale expulsion and "self-deportation" of untold numbers of the nation's best scientific researchers by the Nazi Party in Germany at about the same time. The harnessing of scientific research in the United States for the development of the nuclear weapons that were to be conclusive tools at the end of World War II is also discussed.

Chapter 2 of this book follows this story into the post–World War II era, highlighted by some of the specific issues mentioned in the first paragraph of this preface. The decision as to which events to feature in this chapter was a difficult one, because one of the fundamental truths about the relationship of science and politics is that these two great areas of human thought are in virtually constant conflict in the modern world. One of the messages to be found in the many studies conducted on the politicization of science by governmental and nongovernmental bodies in the 2000s was that the disruption of the scientific process for political reasons can be found almost everywhere in government, in issues involving the environment, research on and the licensing of drugs, development of agricultural policy, testing and labeling of foods and related products, testing and licensing of potentially hazardous facilities (such as nuclear power plants), and endangered species designations and planning. Chapter 3 provides an opportunity for individuals with special interest in and knowledge of the details of issues such as these to present their own viewpoints on them.

This book is designed not only to provide helpful background information on the relationship between science and politics, but also to offer resources and suggestions for further research on this topic. Chapter 4, for example, provides sketches of a number of individuals and organizations with special relationships to the interaction of science and politics, such as the Union for Concerned Scientists, and John Marburger, scientific advisor to President George W. Bush, in office during the time when the most severe criticisms of the politicization of science were being leveled. Chapter 5 is a collection of some data and documents relating to the politicization of science, such as relevant laws, court cases, and position statements on the issue. Chapter 6 is an annotated bibliography of print and electronic resources. Chapter 7 is the chronology of events of significance, and Chapter 8 is a glossary of important terms used in talking about the relationship of science and politics.

Science and Political Controversy

And Xochimilco declares:

That on the fifth hour of the first new moon in the time of Tocoztontli, the snake god Quetzalcoatl will appear on the steps of the Great Pyramid, and

That will signify that Centeotl, the goddess of maize, is ready to receive your seeds and your plants, to sustain you for the new year, and

That, to honor Her, the time is right to give under the knife the firstborn child of one family of the Teotihuacans.

Science—or something like it—has been a part of every human civilization of which we know. Aztecs priests studied the patterns of stars in the sky and the movement of the planets to predict the seasons of the year and, hence, the time for planting. They then built precise mechanisms for determining when important events, such as the arrival of spring, were to occur. The design of pyramids in which the shape of a snake appeared on the steps on only one day of the year was such a device. They also recommended the human sacrifices needed to ensure that otherwise unpredictable natural events, such as the sprouting of seeds and the growth of plants, would continue. (See, for example, Selin 2000.)

The role of science in the life of ancient cultures was very much the same among the Aztecs of Mesoamerica, the early

---

The Trial of Galileo, 1633. (The Bridgeman Art Library)

Chinese dynasties, states of the Indian subcontinent, the Mesopotamian empires, and other ancient civilizations. (We use the word *science* advisedly at this point as the activities of these early scholars do not fit any modern definition of the word *scientist*.) In ancient Babylonia (which dates from about 1894 BCE to about 911 BCE) and other ancient cultures, science was very much the handmaiden of society. Its activities were largely devoted to solving the problems of everyday lives, such as finding materials with which to build structures and techniques for raising those structures, discovering methods for curing disease, establishing the times at which crops should be planted, developing methods for surveying land, and developing a system of counting and currency that could be used in the society's economic transactions. In this regard, Babylonian science reflected a fundamental theme about the goals of science that was to be reiterated time and time again throughout history, even to the present day, that a primary goal of scientific research was to meet the needs of society, to solve the everyday problems that people face in their lives.

An alternative view of science developed in ancient Greece after about 700 BCE. On the one hand, Greek scholars continued to pursue the type of research that had long existed, the effort to understand "how things work" and how changes come about (which the Greeks called *techne*, a term from which the modern world *technology* is derived). Out of this research came a host of inventions still useful in everyday life, such as the gear, screw, rotary mill, screw press, water clock and water organ, the steam engine, the first maps, cranes, caliper, tumbler lock, winch, wheelbarrow, and canal lock. (See *Ancient Greek Technology* 2000; Brumbaugh 1966; Kotsanas 2013.)

At the same time, many Greek scholars begin a new line of research, an attempt to understand nature, regardless of any practical use that research might have. The Greeks called this form of research *epistime*, from the Greek word that means, simply, "knowledge." Perhaps the greatest achievement of this line of research was the development of a new field of abstract mathematics that had little or nothing to do with the problems of everyday life (although such applications did eventually appear in many instances). Some leading figures in this field were Pythagoras (ca. 582–ca. 507 BCE), Euclid (ca. 323–ca. 283 BCE), Archimedes (ca. 287–212 BCE), and Hipparchus (ca. 190–ca. 120 BCE). Of almost equal significance was the birth of a new field of astronomy, devoted to a detailed analysis of the movement, size, distance, and other characteristics of heavenly bodies, with sometimes very complex explanations about those properties. Unlike earlier scholars, the Greek researchers did not study the sky primarily to predict important events on Earth, but simply to better understand the nature of the universe. Among the most prominent of Greek astronomers were Thales (ca. 624–ca. 546 BCE), Eudoxus of Cnidus (ca. 410–ca. 355 BCE), Aristarchus of Samos (ca. 310–ca. 230 BCE), Eratosthenes of Cyrene (ca. 276–ca. 195 BCE), and Apollonius of Perga (ca. 262–ca. 190 BCE). (For more on ancient Greek science, see Clagett 1971; Freely 2012; Kotsanas 2013.)

Again, because of the significant differences in the way the ancient Greeks attacked the objects of their study, it is probably inappropriate to refer to them as *scientists*. More commonly, they are described as *natural philosophers*, scholars who used mental skills such as those of logic and reason to try to make sense out of the observations they made of the natural world.

The success of Ionian science (the name given to the new field of study that began in that part of Greece in about the sixth century BCE) led to the first conflict between science and politics of which we have some evidence from written history. According to historian of science Benjamin Farrington, the problem arose because Ionian science provided humans with a new way of thinking about the world and new explanations about the nature of reality. At the time, he writes, Greece was ostensibly moving toward a democratic political system, although for all practical purposes, it remained an oligarchy. The new science posed no threat to society as long as it was understood only by the ruling class and did not filter down into the lower classes, where it would have a tendency to disrupt the social order. But once it began to appeal to the lower classes, the potential for social unrest became apparent. (For details of this argument, see Farrington 1939, passim.)

In classic Greek society, as in essentially all ancient (and perhaps all or most modern) societies, the masses were kept under control of the ruling classes by imposition of what has been called *The Noble Lie*, a description as to how human society was first created and how it is "supposed" to function (Schofield 2007). The Noble Lie, most clearly stated at the time in Plato's *Republic*, justified the fact that some individuals in a society are destined to be better off than others; some are destined to be rulers and others, the ruled; and that a life of labor and perhaps misery is part of some creator's overall scheme of things for the survival of society. A long tradition of mythological, religious, social, literary, and other types of stories were an essential part of maintaining the Noble Lie. The rise of Ionian science, with its potential for offering ordinary citizens a nonsuperstitious interpretation of human culture different from that presented by the Noble Lie was, therefore, a risk to the social stability of the Greek state.

After a long period of suppression of Ionian science by Greek authorities, the philosopher Epicurus (341–270 BCE) appeared on the scene with the goal of restoring the glory of the embattled Ionian philosophy. He experienced the same resistance from government authorities as had his Ionian predecessors and, as power shifted from Greece to Rome at the end of the pre-Christian era, his followers also met with harassment, abuse, banishment, and, ultimately, failure of their cause. Farrington concludes his analysis of the conflict between Ionian science and Greek authorities by suggesting that this period in history was unusual because it is "not so common to find any corresponding sense of the responsibility for the existence of such ignorance [the superstitions of the Greek populace]; still less of the active part played by governments in the promotion of ignorance" (Farrington 1939, 24), although one must question that judgment in the light of more recent histories of the interaction between politics and science.

Greek Science in Decline

By the first century CE, Greek science had gone into decline, with the revolutionary mode of thinking displayed by scholars such as Thales, Eudoxus, Aristarchus, Eratosthenes, and their colleagues largely having disappeared. Some historians point to the rise of the Christian religion as a major factor in this change, although considerable difference of opinion exists on this point. It seems clear that early Christian philosophers and theologians believed that they were living in the "last days," with the return of Jesus an imminent event that they were likely to witness in their own lifetimes. According to a number of Biblical references, Jesus himself encouraged this belief. For example, in Matthew 10:23, he advised his disciples that "When they persecute you in one town, flee to the next, for truly, I say to you, you will not have gone through all the towns of Israel before the Son of Man comes."

Under such circumstances, the only type of learning really worthwhile at the time was a better understanding of Christian theology that would prepare people for death and entry into the next world. The study of secular topics, such as art, craft, literature, and science, was simply a waste of time. The early Christian author Tertullian (Quintus Septimius Florens Tertullianus) expressed this view in his book *De praescnptione haereticorum* (*On the Rule of Heretics*), where he said that

> We have no need of speculative inquiry after we have known Christ Jesus; nor of search for the Truth after we have received the Gospel. When we become believers, we have no desire to believe anything besides; for the first article of our belief is that there is nothing besides which we ought to believe. (Bindley 1914, 46)

This view is repeated by a number of Church Fathers in the first few centuries after Christ's death. In his 420 CE work, *Enchiridion* (*Faith, Hope, and Love*), Saint Augustine chastises those who waste their time on the study of motion, the location of the stars, plant and animal life, eclipses, and other such natural phenomena, for, he says,

> It is enough for the Christian to believe that the only cause of all created things, whether heavenly or earthly, whether visible or invisible, is the goodness of the Creator, the one true God; and that nothing exists but Himself that does not derive its existence from Him. (Saint Augustine 2013, 7. For more detail on this point, see Clagett 1971, Chapter 10; "Early Christian Attitudes to Science" 2013.)

Whatever the influence that Christian theology may or may not have had on scientific thought, the fact is that natural philosophy was largely moribund in the period from about 100 CE to well into the 16th century. At the end of that period, wherever science was taught, it was still the best known Greek scholars, such as Aristotle, Ptolemy, and Hippocrates, who were accepted as authorities in their field. Little new knowledge was developed through original research or logic. Indeed, were it not for the Islamic world, even more of the knowledge developed during the Golden Age of Greek science would have died out.

One of the great good fortunes in the history of science was the interest expressed by members of the young and vibrant Muslim religion toward secular knowledge in the first millennium of the so-called Christian era. A host of scholars that included the alchemist Jabir ibn Hayyan, the polymath Ibn Ishaq al-Kindi, the physician Hunayn ibn Ishaq, the inventor Abbas ibn Firnas, the mathematician and astronomer al-Khwarizmi, the astronomer al-Battani, the physician Abu Bakr Zakariya al-Razi, the polymath ibn Sina (better known as Avicenna), the astronomer ibn al-Haytham (also known as Alhazen), the poet and mathematician Omar Khayyam, the physician ibn al-Nafis, and the mathematician and astronomer Nasir al-Din al-Tusi learned about Greek texts in natural philosophy, translated them into Arabic, and used them as the basis for a range of new experimental research.

Historians have long argued over the long-term significance of the 1,000-year dominance of Islamic science. Some have argued that Islamic scholars served an important role in preserving and protecting knowledge produced by the Greeks, but contributed little or nothing new to the body of scientific knowledge. Others have suggested that Islamic natural philosophers made many important new contributions to the field and may have laid the groundwork for the scientific revolution that eventually took place beginning at the end of the 16th century. (See, for example, Al-Hassan and Hill 1988; Dallal 2010; Nasr 1976; Saliba 2007.) Whatever the case, Greek knowledge was collected, preserved, and augmented by Islamic scholars and then retransmitted to Christian Western Europe when Muslims invaded and conquered the Iberian peninsula in the eighth century CE and then ruled the area for the next seven centuries.

The Rise of Modern Science

By the end of the 14th century, a movement was underway in Europe in reaction to the suppression of learning that had begun more than a millennium earlier with the rise of Christian theology described in previous section. Slowly, scholars began to display a renewed interest—a Renaissance, as the period was to come to be known—in the study of virtually every field of the human intellect, including art, music, literature, philosophy, politics, and science. For the first time in centuries, scholars began to question the knowledge and methods that had dominated thought for more than 10 centuries. In the field of science, one of the first individuals to adopt this new outlook on study was the Polish scholar Mikolaj Kopernik (Nicholas Copernicus in its Anglicized form). Copernicus was a rather remarkable scholar with a degree in law and interests and aptitude in a wide variety of fields including medicine, art, politics, economics, linguistics, and astronomy. He is best remembered today for his contributions in the last of these fields, specifically in his development of a heliocentric theory of the universe.

Throughout history, the dominant theory of astronomy was the geocentric theory, one that placed the Earth at the center of the universe, with stars, planets, comets, and all other astronomical objects in orbit around it. During the first 15 centuries of the Christian era, this theory was accepted by both scientists and nonscientists alike, not only for empirical reasons, but also because it confirmed the teachings of Christian theology. As early as 1514, however, Copernicus began to have doubts about the validity of the theory. His observations and calculations convinced him that a model of the universe in which the Sun, rather than the Earth, was at the center of the solar system made more sense and better fit the observations that astronomers had been making for centuries. Copernicus was reluctant to publish his theory, however, as he was well aware of the objections that authorities would have to the displacement of the Earth from the center of the universe. That response would come at least in part from the leaders of the Christian church and in part from civil authorities which, at the time and given the influence of the church in state affairs, was essentially the same. Copernicus decided to withhold publication of his book about the heliocentric theory, *De revolutionibus orbium coelestium* (*On the Revolutions of the Celestial Spheres*) until 1543, just prior to his death, thereby avoiding any condemnation and punishment on his "heretical" ideas from church authorities.

Probably no event is better known nor more frequently cited with regard to the conflict between science and religion than is the case involving the inquisition of Italian scholar Galileo Galilei with regard to his views about the structure of the universe. As noted in the previous para, the dominant theory of the universe throughout history had always been a geocentric model. That theory was bolstered not only by hundreds of years of scientific observations, but also by a number of Biblical texts attesting to the idea that everything in the universe revolves around the Earth. For example, 1 Chronicles 16:30 says that "the world is established; it shall never be moved," a point reaffirmed in Psalms 93:1 and in Psalms 96:10, which use exactly the same words, and in Psalms 104:5, which uses the slightly different variant, "[the Lord] set the earth on its foundations, so that it should never be moved" (English Standard Version; http://www.biblegateway.com/passage/?search=Psalm%20 104:5;&version=ESV).

By the early 17th century, Galileo had begun to doubt the truth of the geocentric model. His first concerns about the model arose out of observations he had made with his new handmade telescope, the first (or one of the first) instruments of its kind in human history. With the telescope, Galileo was able to observe the phases of the planet Venus, the moons of Jupiter, and other phenomena that were not consistent with a geocentric universe. Aware of the negative reception his results were likely to receive among church authorities, Galileo was cautious in announcing his acceptance of a heliocentric theory of the universe over the traditional geocentric model. Some influential church supporters were willing to allow him latitude as long as Galileo agreed that heliocentrism was only a theory that, for Biblical reasons, could not actually be true. But his pronouncements on the topic continued to trouble other authorities, and he was summoned to Rome in 1633 to appear before a court of inquisition about his views on geocentrism and heliocentrism.

After studying Galileo's writings on the topic, the court found Galileo guilty of heresy. The heliocentric view was, the court declared, "foolish and absurd in philosophy, and formally heretical since it explicitly contradicts in many places the sense of Holy Scripture" (Finocchiaro 2008, 146). As punishment, Galileo was instructed by the presiding judge, Cardinal Bellarmine, "to abstain completely from teaching or defending this doctrine and opinion or from discussing it; and further, if he should not acquiesce, he is to be imprisoned" (Finocchiaro 2008, 147). In fact, Galileo was never imprisoned, although he did spend the rest of his life (26 years) under house arrest.

The trial of Galileo is usually presented as a perhaps somewhat simple conflict of views between a traditional theological view of the world which could countenance no contrary views against a revolutionary new collection of scientific data which quite obviously provided just such a contrary view. And his trial certainly contained well and enough of that conflict. But it also contained an issue just as profound and just as broad as the question of science versus theology. For Galileo's crime was not just that he was posing a new theory of astronomy, but that he was presenting it in a way that challenged the basic structure of contemporary society.

Had Galileo been willing to confine a discussion of his theories to his academic colleagues, including those within the church itself, he almost certainly would have escaped an inquisition and a trial. But, like his Ionian predecessors, he saw the new discoveries and, more important, the intellectual methods by which they were produced, as a powerful new tool for the general populace, a way of weaning them from superstitious nonsense about the natural world. And he wanted to get word of this new tool out to ordinary citizens *now*, not after a gradual evolution in thought that might take many generations. As Giorgio de Santillana has written in his masterful *The Crime of Galileo*, he

> wanted the scientific awareness spread to the whole advancing front of his civilization, from its expressive and technological capacities and its critical activity to its philosophical reflection. (De Santillana 1955, 205)

And, for that reason, authorities could only see him as a "dangerous novelty-monger" and a "troublemaker" (De Santillana 1955, 205). He posed a threat not only to the Roman Catholic view of the world, but to the whole political structure of society.

And for that reason, he earned the enmity and the censure of not only his long-time foes within the church and the academic community (of which he had many), but also virtually all of his friends and supporters, including Pope Urban VIII, who had pronounced his guilt and his sentence even though he, as Cardinal Maffeo Barberini, had been one of Galileo's staunchest protectors and advocates. (De Santillana's book is essential reading for anyone who wishes to understand the details of the conflict between science and politics during the Galileo trials.)

Galileo's case is important in the history of science (and to readers of this book) because it demonstrates the ability of individuals in authority to "lay down the law" to scientists. The inquisition pointed out to Galileo what he could and could not believe, what he could and could not teach, what he could and could not advocate for, no matter what the data from his research seemed to say. The Galileo case was a warning sign for all future generations of scientists that challenges might occur to their research if that research contradicted or impeded the views of individuals and organizations in power, whether religious groups or secular governments (which, in the case of the Galileo trial, were essentially identical for all practical purposes).

What Is Science?

The Roman Catholic church's action against Galileo came at a crucial moment in the history of science because it was a moment at which a number of scholars were attempting to define for the first time in history what this new intellectual discipline called *science* was to be like. What were its objectives to be? Its methods? Its limitations? Its place (if any) in society? Galileo himself was one of the most important and most influential contributors to this discussion. In his professional life, he was concerned not only about experiments through which he could learn about the natural world, but also about the "rules of the game" under which research was to be conducted, that would ensure that information obtained in these experiments would be valid (providing answer as close to truth as possible) and reliable (providing answers that are consistent from time to time).

Galileo outlined his ideas as to what a "scientific method" should be like in a number of his writings, most importantly *Il Saggiatore* (*The Assayer* 1623) and *Discorsi e dimostrazioni matematiche, intorno à due nuove scienze* (*Discourses and Mathematical Demonstrations Relating to Two New Sciences* 1638). The latter book was written at the end of his life and summarized much of what Galileo had come to believe was essential to the ways of doing science. (For good reviews of Galileo's views on the proper methods for doing science, see Gower 2005 and Mays 1974.)

Over the centuries, the so-called scientific method has been widely promoted as one effective way of studying the natural world. The term, however, has been used in a variety of ways by different individuals, and it is not always clear what a person means when she or he refers to the scientific method. Some of the most important elements of a scientific way of tackling problems, however, are the following:

- Any object or event that can be observed with one of the human senses is a legitimate object of scientific research; by contrast, anything that cannot be studied in this way is not a legitimate object of scientific research.

- Objects and events must be studied in such a way as to produce quantifiable (numerical) results.

- Experiments in science are motivated by some known piece of information about an object or event (a "fact") or by someone's guess about the nature of an object or event (a "hypothesis").

- No statement about an object or event is ever considered to be "true" in an absolute sense. Every such statement is always subject to continued experimentation, modification, further confirmation, or disproof. One can only be more or less certain about the truth of a scientific "fact," never completely certain.

- One's confidence in the "truth" of a scientific statement depends to a considerable extent on the number of times that statement has been tested successfully; the greater the number of experimental replications, the greater confidence one can have in the "truth" of that statement.

- Any experiment must be conducted in such a way that it can be repeated (replicated) by other scientists with training in the same field.

- Efforts to make generalized statements about a set of related facts can result in the development of a *theory* about a phenomenon. In science, the word *theory* normally does not mean a "guess," as it does in everyday life, but a general statement that describes a particular phenomenon, such as the atomic theory of the structure of matter.

- A very large collection of facts and theories is called a *scientific law*. In science, a law is not prescriptive (telling how one *must* behave), but descriptive (telling how nature *does* behave).

A seemingly endless number of descriptions of the scientific method are available in books, magazines, web sites, and other sources. They differ in manner of presentation, completeness, consistency, and other traits. Readers should be familiar with at least some of these descriptions because the conflict between science and politics—both historically and in the modern world—often arise simply because individuals may not understand what it is that science *can* do, how scientists go about their research, what the limitations are under which they work, what the purpose of scientific research is, how the results of scientific research can be put to use in daily life, and so on. (For more detailed discussions of the scientific method, see Gower 2005; Harris 2013; Tibbetts 2013.)

The Organization of Science

The notion of a particular way of doing scientific research (the scientific method) slowly evolved in the years following the late 16th century. As more and more scientists agreed as to how research ought to be conducted, they found that new institutions and organizations were needed to carry out their work. One such institution was the scientific society, a group of individuals with common interests who met to share their ideas and the results of their experiments. The first such society was the Accademei dei Lincei, founded in 1603 by Cardinal Federico Angelo Cesi and named for the lynx, whose sharp eyes represented the power of observation in scientific research. Like many of its successors, the academy included members from a variety of nations, including Germany, Greece, and the Netherlands, as well as Italy. In 1611, Galileo was elected to membership in the society and quickly became its intellectual center. Not much later, two of the oldest, most famous, and most highly esteemed scientific societies were also established, the Royal Society of London in 1660 and the Académie des Sciences in Paris in 1666.

Another new institution that arose in the early 17th century was the scientific journal, a magazine-type publication designed primarily to publish the results of research carried out, often by the members of a specific society. The first such journal may well have been a publication of the Accademei dei Lincei called *Gesta Lynceorum*, although the document apparently never appeared in printed form and existed only as individual papers written by members of the society ("Gesta Lynceorum" 2013). The first scientific journals to appear in print were the *Journal des sçavans*, first published by the Académie des Sciences in January 1665, and the *Philosophical Transactions of the Royal Society*, whose first issue appeared two months later.

Changes such as these reflected an important feature of modern science, a growing sense of community among researchers in a common field. During the time of Copernicus and Galileo, and for many years thereafter, scientific research was largely a solitary activity carried out by a scholar with the assistance of only poorly trained helpers. Over the years and decades, however, that pattern slowly began to change with researchers beginning to work together in teams on their peers on projects of common interest. The ultimate result of that trend can be seen today in modern scientific papers that often listen dozens or even hundreds of coauthors, individuals who have contributed in one way or another to the completion of an experiment. Two such examples from recent history are papers describing the results of experiments on the human genome and on the Higgs boson, each of which listed more than a thousand coauthors (Chatrchyan et al. 2012; International Human Genome Sequencing Consortium 2006).

What Is the Purpose of Science?

The rise of modern science almost inevitably prompted another related field of debate: What should be the primary goal of this new way of studying nature? That is, what ought the purpose of science be? One of the first scholars to deal with this question in detail was the English jurist, statesman, philosopher, and scientist Sir Francis Bacon. Bacon's contribution to the development of modern science has been the topic of considerable debate over the years, but there is no doubt that he expressed a view of modern science that had probably never been so clearly stated prior to his time. Throughout his life, Bacon argued for a study of science that would have a number of features. It would, first of all, increase the general knowledge that humans have about the natural world, whether it had any practical value or not. One of Bacon's most famous contributions was his description of a fictional "Salomon's House," a type of research institute or college that was "dedicated to the study of the works and creatures of God." Scholars employed at the institute spent their time on basic research about nature, accumulating knowledge without regard to its possible applications in everyday life (Bacon 1627).

Scientific research was also a person's mission as a Christian, as it allowed him or her to have a better understanding of God's creative powers. Most important of all, however, was the use of the products of scientific research to produce new devices that would improve the life of all humans. That is, the main goal of science was utilitarian, rather than theoretical. In his treatise entitled *The Great Insaturation*, Bacon wrote that the primary goal of science is to produce "a line and race of inventions that may in some degree subdue and overcome the necessities and miseries of humanity," a topic that he treats in detail in the second part of this work (Bacon 1620; for more detail, see Rossi 1957, 1968). Bacon also attacked traditional methods of learning about the natural world that relied primarily on logic and reason, and insisted that knowledge could be gained only by observing and experimenting with nature directly.

Other scholars also focused on these four primary objectives of modern science: emphasizing the glorification of God through the study of nature; developing practical tools and devices to make human life safer and more enjoyable; accumulating a store of basic knowledge about nature, whether it had utilitarian value or not; and promoting a new way of thinking about and studying nature (the scientific method). In France, the philosopher René Descartes was the primary spokesperson for the new discipline of science, arguing very much along the lines of Bacon in his most famous work *Discours de la méthode* (*Discourse on the Method*), published in 1637. In his treatise, Descartes was very optimistic about the ability of scientific thinking to bring about a host of changes that would transform human life. At one point in the *Discours*, for example, he says that continuing improvements in our understanding of health, illness, and medicines will probably lead to a time when humans will be able to live forever because physicians have found ways to prevent all those circumstances that lead to death (Descartes 1637, Part Six).

Before long, the discussion as to what the goals of scientific research ought to be was overtaken by profound changes in the economies and political structures of many regions of Western Europe, most notably the Industrial Revolution that began primarily in Great Britain and the French Revolution of the late 18th century.

The Industrial Revolution

The Industrial Revolution is an era in history during which profound changes occurred in a number of fields, including transportation, mining, agriculture, the textile industry, and the chemical industry. A list of the new inventions produced during this period is truly impressive and includes items such as the steam engine (along with its many applications in transportation); flying shuttle used in weaving; spinning jenny, spinning frame, and spinning mule; power loom; telegraph and telephone; sewing machine; screw propeller; mechanical reaper; combine harvester; iron buildings and bridges; arc lamp; steam-powered rock drill; radio; and linotype. (A complete list of inventions requires more space than is available here. See the excellent history of the period at Derry and Williams 1993 and a partial list of inventions at "List of Inventions and Inventors During the Industrial Revolution (1800–1899)" 2013.) How was this explosion of inventions associated with the growing field of science discussed up to this point in this book?

In today's world, it is possible to distinguish among three types of related activities that can be labeled *pure science*, *applied science*, and *technology*. Pure science is research that is conducted solely because of someone's interest in a particular question about the natural world. For example, thousands of scientists have spent billions of dollars working over many decades to find out what matter is composed of. The most recent breakthrough in that research occurred in 2012 when a very large research team using the multibillion-dollar Large Hadron Collider discovered the Higgs boson, a particle that apparently gives mass to matter. The project is designated as pure science because it was conducted simply to learn more about the nature of matter, not to solve any specific problem in everyday life.

By contrast, applied research is conducted for just that purpose, trying to discover the solution to some problem in the everyday world. When researchers announce the discovery of a new way of treating a disease, they are reporting a breakthrough in applied science. They have used science as a way of dealing with a specific medical problem.

Some people equate technology with applied research. But others point out differences between the two endeavors, the most important of which may be the background from which one attacks a problem. An inventor might use scientific principles in developing a new device, for example, but not necessarily have any special training in science itself. (One interesting analysis of this complex issue is Jones 2013. See also Feibleman 1961.)

According to these definitions, the vast majority of inventions made during the Industrial Revolution were technological advances, not developments in science. Most of the inventors responsible for these new devices had little or no scientific training; they were much more likely to have come from an arts and crafts background or, perhaps, with no technical training at all. For example, Thomas Newcomen, who invented the first steam engine, was an ironmonger and a lay preacher for the Baptist church. John Kay, who invented the flying shuttle, had worked for a time on hand looms, but had no other technical training or career. James Hargreaves was a carpenter and a weaver who invented the spinning jenny to vastly improve the efficiency of a craft he knew well. James Watt, perhaps the most famous name of all in the Industrial Revolution, was working as an apprentice instrument maker in Glasgow when he began designing improvements on Newcoment's steam engine. Clearly, none of these individuals (nor many of their peers) were products of the new field of science created by Copernicus, Galileo, Bacon, and other famous names in science.

The most important feature of the Industrial Revolution for the growth of science, then, was that it created for the first time in history a new class of workers whose focus was not on the creation of knowledge, per se, but the attack on very specific problems in everyday life, whether they be called advances in applied science or developments in technology. What is important to the scientific endeavor is that these changes opened up a whole new view of what science ought to be like. One change resulting from the Industrial Revolution was the creation of new institutions of learning for young inventors and their peers who could never have been admitted to traditional universities. Among these in England were a number of colleges of arts and science, colleges of mechanical arts, and commercial and mathematical schools designed specifically for working-class men (and it was essentially always men) who wanted to learn more about science, but who also had to hold down regular day jobs. For this reason, many classes at the new institutions were held in the evening.

Another change was the establishment of scientific societies broadly similar to the Royal Society, but open to lower and middle class men who could never have been admitted to that organization. These organizations provided an opportunity for individuals of the lower and middle class to attend classes, hear lectures, carry out experiments, and present the results of their own research. An example of such a society was one formed in Manchester in 1781 as a way of exposing working-class men to the scientific knowledge available at the time. The society provided a variety of forms of instruction for members, published reports by its members in its *Memoirs*, and eventually founded its own college of arts and sciences (Evans 2013; for an interesting review of one of the most famous, although short-lived, societies, see Schofield 1957).

The rise of technology in England during the eighteenth and early 19th centuries along the lines outlined here was accompanied by a corresponding decline in the more traditional forms of pure science practiced by Galileo and his successors. In fact, members of the traditional scientific community began to express concerns that their own approach to scientific research was dying out in the face of the spectacular successes of technological developments. Perhaps the best known expression of this view was a book published in 1830 by the great English mathematician Charles Babbage, *Reflections on the Decline of Science in England*. In his book, Babbage used the case of the Royal Society to show that traditional scientific research (in contrast to technological developments) was in a state of decay in the nation. The government no longer provided adequate financial support to the institution; it had lost the respect of both the government and the general public; its members were largely lay people, rather than scientists, who published little scientific research of any consequence; and the society itself apparently no longer knew what its role in modern society was to be (Babbage 1830).

The irony of the role of science and technology during the Industrial Revolution, then, was that technological developments were occurring at a rate unprecedented in human history, producing social, economic, political, and other changes never seen before in human history. But at the same time, professionals scientists interested in pure research were complaining that their status and work were diminishing the depths they had never expected to see.

The French Revolution

A democratization of science similar to that which occurred in England during the Industrial Revolution also took place in France during the revolution of 1787. The impetus for the scientific changes in France were, however, very different from those in England. The message behind the French Revolution was, of course, the overthrow of the monarchy and the aristocracy and the transfer of power to ordinary citizens. As an adjunct to that movement, the leaders of the revolution turned to science as a critical element in the reimagination of French society. The first step in that process was the retreat from the concept of science simply as a way of learning more about the natural world, a philosophy that had made France the world's leader in traditional science in the seventeenth and eighteenth centuries. Instead, the goal of science had to be changed to focus on the needs of French citizens, the improvement and advancement of French society.

To that end, many revolutionary reformers placed instruction in science at the forefront of the new educational system developed following the revolution. A major spokesperson for this position was Marie Jean Antoine Nicolas de Caritat, the Marquis de Condorcet, who was himself an eminent mathematician, philosopher, and political scientist. Condorcet avidly believed that education in the sciences was the best training that a person could have to become a careful, analytical thinker and, therefore, a good citizen. He was convinced that a citizenry trained in thinking like a scientist and using the scientific method to deal with all types of problems would be able to contribute to the development of the new French nation that was then being developed. In his outline of a proposed educational system for the new republic, Condorcet called for a five-tier system of public education: primary, secondary, institutes, lycée, and a National Society of Science and the Arts. The focus on training in science was to increase at every level until it was by far the dominant activity at the upper three levels of the system (Condorcet 1791; see also Chafy 1997).

Virtually all traditional scientists fell in with the revolutionary movement, adding their expertise to the process of dealing with the myriad of issues involved with building a new society. There is no way of knowing the extent to which these scholars fully and truly adopted the philosophy, goals, and methods of the revolution and which became part of the movement as an act of self-preservation. The movement was itself strongly anti-establishment and before it had collapsed, it had destroyed virtually all of the traditional scientific establishments in France. Indeed, even those scientists who joined the revolutionary movement did not, by doing so, necessarily save their own lives. For example, the greatest French chemist of the day, Antoine Lavoisier, was an active participant in the early days of the revolution, but eventually lost his head to the guillotine in 1794 (Burns 2003, 105–6; Gillispie 2004).

The hopes and expectations that many people had for converting French society into a modern, rational state as a result of the revolution were only partially realized. For a number of reasons, plans for basing a social order on the logical process of scientific thinking failed to a considerable extent. Although effective at destroying traditional scientific establishments to which they objected, French revolutionaries were not as successful in building new institutions or in harnessing the energy of scientific research for social purposes. The one exception to that fact was the creation of a new metric system in 1795 by the Commission of Weights and Measures (which included such prominent scientists as Condorcet, Pierre-Simon Laplace, Adrien-Marie Legendre, and Lavoisier). That system, in a slightly revised form, is now the system of measurement used by all scientists in the world and all but a very small handful of nations for everyday measurements (Alder 2002).

The Origins of Science in America

Science in Colonial America was very different from its European counterpart in one important way: It had no long tradition of basic research like that embodied in the works of Galileo, Bacon, Descartes, Newton, Huygens, and dozens of other great scientists. American science did, however, reflect many of the goals of 17th- and 18th-century British science, at least partly because, after all, the American colonies were *colonies*, peopled by men and women from the homeland. Immigrants to the New World brought with them many of the same beliefs held in England with regard to science, namely that a major goal of science was to make life better for citizens of the society. This challenge was even more important in the colonies because of the very significant struggle most immigrants faced to simply stay alive and prosper in this new land. It is hardly surprising, then, that much of the research done in North America up until the 1800s focused on exploring the natural resources of the new continent and finding ways to use these resources for the betterment of residents of the colonies. Topics in which American scientists were interested included primarily utilitarian arts, such as surveying and mapmaking, mining and processing of metals, agriculture, navigation, fishing, construction, and military arts and crafts. At this stage in history, science was an entirely democratic pastime because anyone could make a contribution to the advancement of knowledge, by going into the fields and finding and identifying minerals, by discovering the available plants and animals, by collecting data on tides and other ocean movements, and by studying the movement of stars and planets and producing new astronomical data (Burns 2005).

No more than a handful of early colonists were interested in the more abstract aspects of science, and these few scholars either returned to live in England or spent much of their lives there. The most famous of these individuals was Benjamin Franklin, better known today for his political activities than his scientific achievements, but an internationally recognized scholar in science during his lifetime. In addition to a host of practical inventions, such as bifocals, the Franklin stove, an odometer, and a musical instrument known as the harmonica, Franklin conducted a number of original experiments, primarily in the field of electricity, and contributed to the theoretical debate over the nature of electrical charges (Cohen 1956).

Until the last quarter of the 19th century, the United States produced no more than a handful of world-class pure scientists other than Franklin, the most famous being Joseph Henry. Henry made a number of classic discoveries in the fields of electricity and magnetism, although his work was largely ignored by his European counterparts. For the most part, they were willing to have American researchers focus on the study of their natural environment and the production of utilitarian devices that found use both within and outside of the New World. In his remarkable book on the history of the American government's attitude toward basic research, Daniel Greenberg points out that "[p]rior to World War II, . . . science existed more or less as an orphan on the American landscape. . . . [Congress] was frequently willing, sometimes eager, to employ public funds and authority to apply knowledge for public purposes, but it yielded only slowly in its belief that public funds should not be used to support the acquisition of knowledge" (Greenberg 1967, 52, 56; see all of Chapter 3, "When Science Was an Orphan," for an excellent review of the role of basic research in pre–World War II America).

The American government also reflected this view of science, providing essentially no support for basic research and offering financial resources only to utilitarian agencies such as the Army Corps of Engineers, created in 1802; Survey of the Coast (the forerunner of the modern National Oceanic and Atmospheric Administration), established in 1807; Naval Observatory, created in 1830 as the Department of Charts and Instruments; the Corps of Topographical Engineers, established in 1838; and the Geological Survey, established in 1879. In fact, the only U.S. federal agency created primarily to promote the pursuit of science was the Smithsonian Institution, established in 1846 with funds bequeathed to the United States by British scientist James Smithson in 1829. Henry was appointed first secretary of the Smithsonian and decided that utilitarian research was already well funded in the United States, and that the Smithsonian should focus its attention on basic research (Price 1954, Chapter 1; Reingold 2008; also see Dupree 1957, Chronology).

The Growth of Basic Research in the United States

The use of scientific research for utilitarian purposes was a dominant theme of American science throughout the 19th century and into the early 20th century. A list of projects for which federal appropriations were granted reflects this trend, including authorizations for research on a system of weights and measurements (1830); support for the Morse telegraph

(1842); surveys of federally owned lands (1847); expeditions to the Arctic, the Dead Sea, Chile, and other locations (various dates); establishment of the Department of Agriculture (1862); creation of the national Fish Commission (1871); creation of the Bureau of Animal Industry in the Department of Agriculture (1884); creation of a system of agricultural experiment stations (1887); and founding of the Weather Bureau (1890) (Dupree 1957, Chronology).

By the early 1870s, however, an increasing emphasis on the role of pure science began to occur in the United States. This change was made possible by the creation and development of a number of institutions on which basic research depends in any nation, new institutions of higher learning, new scientific societies, and new forms of support from governmental and nongovernmental organizations. Among the first of these institutions was the Johns Hopkins University, established in 1876 with a grant from Johns Hopkins, an entrepreneur who made a vast fortune through investments, especially in the Baltimore & Ohio (B&O) railroad in the 1840s and 1850s. The new institution devoted itself to fundamental research and teaching in the basic sciences and eventually graduated a number of men (and a few women) who were to become leaders in the new American science of the late nineteenth and early 20th centuries.

Johns Hopkins was representative of the trends in higher education at the time. The first doctorate in any field in the United States was issued in 1861 by Yale University; by 1871, there were only 198 postgraduate students in all institutions of higher learning in the country. At that point, the momentum for graduate education had begun and 20 years later, that number had grown to 2,872 postgraduate students. About 65 percent of the doctoral degrees issued during that period were in some field of science (Metzger 1955, 104; Thurgood, Golladay, and Hill 2006, Tables 2–3, 9).

The last quarter of the 19th century also saw the formation of a number of scientific societies for specialized fields, such as the American Chemical Society (1876), the Biological Society of Washington (1880), the American Geological Society (1891), the American Astronomical Society (1899), and the American Physical Society (1899). (The formation of the American Medical Association had preceded these associations by about a half century, having been formed in 1847.) Prior to this time, only broad, nonspecialized scientific associations had been formed, the most important of which was the American Association for the Advancement of Science (AAAS), founded in 1848. (See Bates 1965 for the best discussion of this topic.)

In nearly all phases of the development of a scientific profession, the United States was about four decades behind Great Britain. By the end of the 19th century, however, the United States was well on the way to developing a resource for the support of scientific research that had never existed to any considerable extent in Great Britain: the availability of very rich individuals who were willing to donate very large amounts of money for the establishment of institutions for the pursuit of basic research. An example of this trend was the philanthropy of Andrew Carnegie in the field of genetics.

Andrew Carnegie was a Scottish-born industrialist who made his fortune in the railway and steel industries in the period between 1860 and 1900. At the peak of his career, his estimated worth was nearly $475 million ($300 billion in 2008 dollars), making him one of the 10 richest Americans who has ever lived ("The Richest Americans" 2013). By 1895, Carnegie had begun to retire from active participation in business and to devote the rest of his life to giving away his personal fortune for charitable purposes. Eventually he endowed nearly two dozen major philanthropic projects, ranging from the Carnegie system of public libraries and Carnegie Mellon University to the Carnegie Endowment for International Peace and the Carnegie Hero Fund. Among those activities was the Carnegie Institution for Science, to which he eventually gave a total of $22 million and whose purpose it would be to "in the broadest and most liberal manner encourage investigation, research, and discovery [and] show the application of knowledge to the improvement of mankind" (Dupree 1957, 297). Carnegie selected as the first president of the Carnegie Institution Daniel Hoyt Gillman, who had previously served as founding president of the Johns Hopkins University. The institution remains one of the nation's most illustrious supporters of basic research in the United States.

One of the first fields in which the institution provided financial support was the Department of Experimental Biology, established in 1905. The purpose of the department was to carry out basic research in biology, for which it created the now-famous Cold Springs Harbor Research Station on Long Island, New York. Research at the station focused to a large extent on genetics, and within a short period of time, that research had projected American biologists to the forefront of the field of genetics worldwide (Micklos 1988; *Year Book—Carnegie Institution of Washington* 1904).

The success of the Carnegie Institution reflected the one advantage that American scientists had over researchers in virtually every other part of the world: access to the largesse of philosophers such as Andrew Carnegie, Andrew Mellon, J. P. Morgan, and John D. Rockefeller, each of whom gave tens or hundreds of millions of dollars for the support of science in their lifetimes. Thus, while researchers in many parts of the world were recovering from the devastation of World War I and the global depression of the early 1930s, American scientists still had access to funds that allowed them to maintain a high standard of basic research.

Science and Politics in the Mid-20th Century

The mid-20th century saw a continuing expansion and growth in the institutions of modern science and an evolution in the variety of ways in which science and politics could interact. This progress varied from country to country, depending on a number of factors, not the least of which was the dominant political philosophy in each nation.

The Soviet Union

In the Soviet Union (The Union of Soviet Socialist Republics [USSR]), for example, the dominant political philosophy changed dramatically as a result of the revolution of 1917, in which the autocratic rule of the czars was overthrown and replaced by a Bolshevik (Communist) government which renamed the nation the Russian Soviet Federative Socialist Republic (RSFSR), more commonly referred to as the Russian Federation, Soviet Russia, or simply Russia. The new Soviet system was based on a socialist philosophy, one in which the central government has control over the production and distribution of all goods and services, with the aim of meeting the immediate needs of citizens rather than the accumulation of wealth by a fraction of the population.

Science was one of the resources over which government was expected to take control, and the Soviet government early on began to designate the ways in which scientific research was to focus primarily on utilitarian functions, designing products and services that would make the lives of the Soviet population healthier, safer, more productive, and more satisfying. The primary institution through which these objectives were to be achieved was the USSR Academy of Sciences, which was first established by Peter the Great in 1724. Under the new Bolshevik government of Vladimir Lenin, the academy was renamed the USSR (or Russian) Academy of Sciences and reorganized in 1918 so as to focus on the specific needs of creating a new socialist state. An understanding was reached between scientists and political leaders that researchers would work on such projects in exchange for whatever funding was needed from the central government ("History of Russian Academy of Sciences" 2013).

For the first few years, Russian scientists had some degree of autonomy in their research, provided they worked within the general constraints laid out in the 1918 transition. By the end of the 1920s, however, members of the Communist Party were becoming frustrated that some scientists were still spending too much time on basic research, not having adequately committed themselves to the party's demand for more utilitarian research. Led both by high officials in the party and by restive revolutionary students, the government initiated a purge of scientists whom it deemed to be too politically unreliable. Eventually, hundreds of working scientists were dismissed or relocated and, in 1930, more than 100 more were brought to trial in an event now known as the Academicians' Case. The government also became proactive, ordering hundreds of dependable young members of the Communist Party to enroll in science and technology programs, with the goal of remaking the scientific establishment in the image of the party itself (Bidlack and Lomagin 2012, 380–382 [a documentary report of the Academicians' Case]; Hirsch 2005, 138–143; Joravsky 1960, 1961 [Chapter 17]).

Probably the best known, and almost certainly the most dramatic, example of the interference of the Russian government in science was the adoption of Lysenkoism as the guiding principle of agricultural practices in the country. Lysenkoism is named after the foremost proponent of this agricultural practice, Trofim Denisovich Lysenko. Lysenko was born to peasant parents in Ukraine in 1898, and he attended the Kiev Agricultural Institute. There he became familiar with and an advocate of the Lamarckian theory of evolution, namely that organisms can pass on to their progeny physical characteristics developed during their lives. He put this theory to a test while working at an agricultural experiment station in Azerbaijan. The experiment he designed involved exposing wheat seeds to very cold, damp conditions, and then seeing how well they germinated and grew compared to seeds exposed to normal environmental conditions. The results of this experiment convinced Lysenko that this process, an extreme example of *vernalization*, not only produced larger crops in the first generation, but transmitted that trait to later generations of the seeds.

Word of Lysenko's results soon reached party headquarters in Moscow, where he was adopted as a savior for the young nation's desperate efforts to increase its agricultural output. To the extent that his results and his theory were correct, the regions in which wheat could be grown would be vastly increased, bringing in the crops the Soviet Union so badly needed. Soviet leader Joseph Stalin brought Lysenko to Moscow, where he was elected a member of the Russian Academy of Sciences in 1935. He was then appointed director of the All-Union Institute of Selection and Genetics in 1936 and chosen president of the All-Union Academy of Agricultural Sciences in the same year. He was also chosen as a member of the Presidium of the Soviet Academy of Sciences in 1938, and appointed director of the academy's Institute of Genetics in 1940. In these positions, Lysenko had virtually a free hand in determining agricultural policy and practice for the Soviet Union into the 1950s.

The only problem with this story is that most biologists in the mid-20th century had long since abandoned the theories of Lamarck, in favor of the genetic theory of evolution that originated in the work of Gregor Mendel at the end of the 19th century. (Lysenko rejected Mendel's findings and theories.) More to the point, the experimental results Lysenko obtained in his Azerbaijan studies turned out to be an artifact; his theory of vernalization was incorrect, and seeds treated with extreme environmental conditions actually performed no better—and in some cases, more poorly—than did seeds planted under traditional conditions. The point of the Lysenko affair is that the science involved was irrelevant to the role played by Lysenko in the guidance of Soviet agricultural. He was an ardent supporter of the Communist revolution and of the goals and methods of the new society, and that was sufficient to win the approval of government officials (especially Stalin) who decided that he and his theories of biology would lead the nation away from its agricultural deficiencies. (After the fall of Stalin, Lysenko fell largely into disrepute, and died out of official favor in 1976.) (Joravsky 1970; Lecourt 1977; Medvedev 1969; Soyfer 1994)

Nazi Germany

Delayed by a period of about two decades, a new government in Germany was going through many of the same trials and tribulations as those occurring in the Soviet Union in the mid-1900s. Battered by defeat in World War I, a crushing reparations penalty, and an international financial collapse, Germans rejected the leadership of the ruling Social Democratic Party in the elections of July 1932, and gave a new majority to the National Socialist German Workers' Party (more commonly known as the Nazis), under the leadership of Adolf Hitler. As in the Soviet Union, the new rulers were faced with the task of rebuilding a struggling nation and, again like the Soviets, they decided to remake the scientific establishment as part of the solution to this problem.

The German solution was somewhat different from the Soviet approach, however. Hitler and his colleagues were convinced that a critical reason for Germany's state of collapse was the influence of the Jewish community in national affairs. This belief was eventually to lead to one of the most horrific examples of genocide in human history, the Holocaust. More specifically and more immediately, however, it also led to the condemnation of a supposed "Jewish science," which the Nazis believed was responsible for a technological collapse in Germany, as well as a corruption of the very process of scientific research in the nation. This problem could be solved, the Nazis believed, only by the extermination of "Jewish science" and its practitioners, and its replacement by a "German science" (also known as an "Aryan science" or "Nordic science" to reflect its ethnic basis) that embodied the ideals of a pure race of Nordic males.

In retrospect, this view was somewhat anachronistic because, during the first decades of the 20th century, Germany had become the world's leader in the natural sciences. One concrete measure of this fact is the number of Nobel Prizes won by German scientists, compared to those won by scientists from other nations. Of the 100 prizes awarded between 1901 and 1932 in physics, chemistry, and physiology of medicine, German scientists earned 23 awards, twice as many as Great Britain (12) and France (12) and six times as many as the United States (4). During the period, scientists from the Soviet Union won only one Nobel Prize ("Create a List" 2013). Strikingly enough, 10 of those Nobel Prize winners (43 percent) were Jewish, a striking contrast to the overall population, in which just over half a million of Germany's 67 million population (about 0.75 percent) were Jewish ("Germany: Jewish Population in 1933" 2013). It would seem to have been obvious that expelling Jews from the scientific establishment in Germany meant essentially gutting the profession and starting over, using only certified members of the Aryan stock. But the Nazi Party determined that, as in Russia, political correctness was of greater importance than scientific aptitude, and the process of the purification of science began.

Nonetheless, prohibitions on the involvement of individuals of Jewish heritage were put into effect very early in the rule of the Nazi Party. In April 1933, for example, only two months after Hitler had come to power, the Reichstag passed the Law for the Restoration of the Professional Civil Service, which outlined requirements for individuals employed as teachers, professors, judges, or other civil servants. Among those requirements was a pure Aryan heritage, which excluded anyone with even a trace of Jewish blood in their veins. For all practical purposes, all Jewish scientists working in academic or government settings (essentially all Jews in the scientific establishment) were required to leave their jobs. (A similar law applying to lawyers, doctors, tax accountants, and other professions was passed shortly thereafter. "Law for the Restoration of the Professional Civil Service, April 7, 1933" 2013.) Additional laws also limited the number of Jewish children who were allowed to attend public schools, eventually reducing that number to essentially zero by the end of 1938 ("Chronology of Jewish Persecution" 2013; "Examples of Anti-Semitic Legislation, 1933–1939" 2013; Mendes-Flohr and Reinharz 2011, Part XI: The Holocaust). As a result of these actions, practicing Jewish scientists were driven from universities, government agencies, and other places of employment; Jewish graduate students were not allowed to finish their degrees; and younger Jewish students were not allowed to attend schools of any kind that would allow them to pursue their interest in science and mathematics.

As a result of these actions, Jewish scientists, by the 1930s, had begun to flee Germany *en masse*. A reflection of that trend was the formation in 1936 of a committee to collect the names of "threatened" German scientists (which, for all practical intents, meant Jewish scientists). The purpose of the list was to start finding destinations outside Germany to which these scientists could be relocated if and when they chose to emigrate. When first published, the list ran 125 pages in length and contained 1,639 names, many of whom were leaders in their field of science, such as Max Bergmann, Hans Bethe, Max Born, Peter Debye, Albert Einstein, Kasimir Fajans, Otto Frisch, Otto Loewi, Lise Meitner, Otto Meyerhof, Gustav Neuberg, Otto Stern, Leó Szilárd, Edward Teller, and Eugene Wigner. A year later, the list was updated with the addition of 154 more names (Emergency Committee in Aid of Displaced German Scholars 1936, 1937; also see "Emergency Committee in Aid of Displaced Foreign Scholars Records, 1927–1949" 2013).

The vast majority of those in the two lists eventually found homes in nations around the world, especially Great Britain and the United States, where they made immeasurable contributions to the development of science and technology in those countries. (See, for example, Moser, Voena, and Waldinger 2013.) By the end of the 1930s, Hitler had achieved his goal of eliminating Jews from the German scientific establishment but, in the process, he had essentially given away one of the greatest collections of scientists the world has ever seen and a vast resource that might conceivably have helped him win World War II (Medawar and Pike 2001).

Great Britain

The drive to use science as a force for social development and improvement extended well beyond the boundaries of the Soviet Union and Germany. During the 1930s, many British scientists also pointed to scientific research as the saving asset in an economy that had plunged to the depths. A number of commentators expressed the belief that science possessed a mechanism (the scientific method) for discovering the truth about the natural world, and that it was their obligation to bring both that method and the fruits of their research to politicians and the government, where it could be used to solve specific social problems.

The seminal event in this period of history was a meeting held in London during June 29–July 4, 1931, under the auspices of the International Congress of the History of Science and Technology. Originally planned primarily as a means of restoring interest in this subject among scientists, the London meeting turned into a revolutionary event when members of the Soviet delegation read a number of papers outlining their view of the role of science in society. (The papers were almost immediately collected and published as a book called *Science at the Crossroads*.) The Soviet authors cited example after example of ways in which science had been harnessed to solve specific practical problems in the new Soviet society, offering these as examples of a Marxist interpretation of the applications of scientific research.

Science at the Crossroads set off a furor within the scientific community in Great Britain, as well as within the political system. Some observers thought the Soviet ideas were well beyond the pale in terms of harnessing science to deal with such a wide variety of social issues. But a number of leading British scientists immediately bought into the Soviet argument, adopting the notion that the primary goal of scientific research out to be utilitarian. The leader of this movement was the eminent Irish crystallographer John Desmond (J. D.) Bernal, who expressed his views on the social role of science in a now-classic work, *The Social Function of Science*. In this book, Bernal argued that science had always been a powerful force in human history, but that its potential was restricted within capitalistic political systems. In order to achieve the maximum that science could contribute to society, Bernal said, a socialist political system must be put into place which could direct the fruits of scientific research (Bernal 1939; Sheehan 1985, Chapter 5). Even late in his life, Bernal was arguing for the powerful potential of science for solving the world's problems. In his 1958 book, *World without War*, he writes that he puts forth the optimistic view of the ability of scientific research to solve social problems. "With the knowledge and experience already at our command," he says, "we could build a world that would provide for every single person in it" (Bernal 1958). All that is lacking, Bernal claimed, is the political will to create the agencies through which scientific knowledge can actually be put to work in this campaign.

Bernal's view did not negate the need for basic research, and a number of his colleagues suggested a dualistic view of the scientific enterprise, in which researchers were isolated (figuratively or literally) in situations where they could carry out their studies without the interference of outside forces, the fruits of which were then put to use by the government for the solution of social problems. As an example, in his presidential address to the British Association for the Advancement of Science in 1933, Sir Frederick Gowland Hopkins suggested that reconsideration might be given to a modern form of Bacon's Solomon House for scientific research. He emphasized that scientists had to be able to work on their own without the influence of outside forces in producing knowledge which could then be put to use in society:

When civilisation is in danger and society in transition might there not be a House recruited from the best intellects in the country with functions similar (mutatis mutandis) to those of Bacon's fancy. A House devoid of politics, concerned rather with synthesising existing knowledge, with a sustained appraisement of the progress of knowledge, and continuous concern with its bearing upon social readjustments. (British Association for the Advancement of Science 1933)

Some scientists went even further in their view of the extent to which science could contribute to the development of science. They argued that because scientists were accustomed to using a rational approach to the solution of problems (the scientific method), they were qualified to deal with a host of social issues outside the immediate purview of science itself. Some researchers who held this view even recommended a transformation of government in which scientists actually made social policy for the nation. Scottish journalist and member of the British Parliament Ritchie Calder at one point, for example, recommended that the House of Lords be disbanded and replaced by a Senate of Scientists that would have legislative authority comparable to that of the House of Commons (Hartley 1963, 39).

Calder was by no means the only proponent of this view. In his 1934 book, *If I Were Dictator*, for example, evolutionary biologist Julian Huxley said that one of his acts as dictator would be to appoint a "Science Council" to advise the government as to the best ways to make use of existing scientific knowledge to deal with the nation's problems. In that way, the existing system would be reversed, in which a governmental system with less rational basis would dictate to scientists what type of research they should carry out. (For more details, see Huxley 1934, Chapter 6.)

Nor was Huxley the most extreme in his views as to the potential role of scientists in directing the course of social policies and actions. In fact, he was only one of a group of scientific prophets who saw a gradually increasing role for scientists in the control of governments, a movement that has been characterized as *scientific dictatorship*. One of the leading theorists of this movement was a man far better known for his works of science fiction, H. G. Wells. Throughout his life, Wells honed a new view of society that was organized around scientific principles and run by scientists. Within this society, individual citizens could not claim the adjective "individual," as they were really only a part of a social machine that operated to produce the best possible world for the greatest number of people. Wells described this new political system in a series of books in which, according to one analyst,

> mankind is shown as progressively emerging from the "decadence," "the twilight of social order" of the mid-twentieth century into a far more exalted state in the years of the 21 first century. This culmination is a technocratic achievement . . . [that is] initiated by groups of scientists and engineers who "invade" politics in a "movement that spreads from workshop to workshop and from laboratory to laboratory" [that resulted in] a world renaissance and the emergence of the World State that Wells believed had become necessary if "the species was not to collapse, degenerate and perish by the wayside." (Rayward 1999)

(For more details about Wells' vision of the future and its consequences for the nature of society, see any one of his books devoted to this topic, such as *Anticipations of the Reaction of Mechanical and Scientific Progress Upon Human Life and Thought* [1902]; *A Modern Utopia* [1905]; *The Open Conspiracy: Blue Prints for a World Revolution* [1928]; *The Shape of Things to Come* [1933]; and *World Brain* [1938].)

The views expressed by Bernal and his colleagues were by no means universal. In fact, as the push for a utilitarian focus of society gained steam, some researchers began to worry about the risks that movement posed to basic research. If the work of scientists is to be directed at solving the problems of everyday life, they asked, what is to happen to the centuries-long tradition of research purely for the sake of learning more about the natural world?

The primary spokesperson for this view was John R. Baker, professor of cytology at the University of Oxford. In a letter to a historian of science Robert E. Filner some years after the 1930s controversy (Filner 2013), Baker explained that he had gradually become more and more agitated at the ideas being proposed by Bernal, and he finally "exploded" in a now-famous article, "Counterblast to Bernalism" (Baker 1939). He said:

> Bernalism is the doctrine of those who profess that the only proper objects of scientific research are to feed people and protect them from the elements, that research workers should be organized in gangs and told what to discover, and that the pursuit of knowledge for its own sake has the same value as the solution of crossword puzzles. . . . What a scientist ought to do is an ethical concern for the judgment of his own conscience. Those to whom one listens with respect when they speak of verifiable matters (e.g., in crystallography *[Bernal's field]*) compel attention much less inevitably when they try to lay down the law on moral issues. (Filner 2013)

One of Baker's major supporters was the great philosopher of science, Michael Polanyi. Polanyi agreed with the general intent of Bernalism, but suggested that it was naive and impossible to achieve in practice. In his 1962 book, *The Republic of Science*, Polanyi looked back on the 1930s controversy and observed that he "appreciate[d] the generous sentiments which actuate the aspiration of guiding the progress of science into socially beneficent channels, but I h[e]ld this aim to be impossible and nonsensical. Any attempt at guiding scientific research towards a purpose other than its own because it is an attempt to deflect it from the advancement of science," he went on. "You can kill or mutilate the advance of science, you cannot shape it. For it can advance only by essentially unpredictable steps, pursuing problems of its own, and the practical benefits of these advances will be incidental and hence doubly unpredictable" (Polanyi 1962, 62).

Baker and Polanyi joined together in 1940 to establish an organization to counteract what they saw as the threat of Bernalism, the Society for Freedom in Science. The goal of the organization was to ensure that scientists had the opportunity to pursuit basic research on topics of interest to themselves, without being forced into working on issues of social importance. They were particularly interested in exploring Lysenkoism and its effects on Soviet science and remained in existence until well after the end of World War II (McGucken 1978; Reinisch 2013).

As vital as the debate was among English scientists in the 1930s about the purpose and organization of science, the discussion was soon swept aside with the advent of World War II. With the outbreak of the war in 1939, and until its conclusion in 1945, no one could argue that scientists should go off on their own to conduct basic research that might or might not contribute to the war effort. The fight for survival demanded the commitment of every trained scholar to victory. The debate about science and politics had to be put off until after the war, when it, in fact, did arise once more, especially in the United States, where it has continued in a variety of forms ever since.

Conclusion

If there is one consistent message that runs through this chapter, it is that science has always and everywhere been the handmaiden of society. Whether the specific task was to predict the arrival of the seasons, find more efficient ways to grow crops, devise new techniques for the processing of metals, build better modes of transportation, or develop more powerful military weapons, science has been called upon to solve the practical problems of governments and social systems throughout the ages. In some cases, these developments have occurred in a relatively random and haphazard way, the consequence of a particular individual's interest in working on some specific problem (such as finding a more efficient way to pump water out of mines). In other cases, governments have taken charge of the scientific activities in a nation to make sure they focus on national priorities. The latter pattern holds true almost everywhere in the world today, with governments supplying financial support for all types of applied science and technology, from the search for the cures for disease to the design and construction of more efficient electronic devices to the production or more powerful military systems.

Basic research has also always been part of the scientific enterprise. Throughout history, there have always been individuals who made observations and conducted experiments simply to learn more about the natural world, whether that knowledge had practical value or not. For the greatest part of history, this form of research was carried out by individuals working in their private laboratories at their own expense. During the 18th century, for example, most of the most important discoveries in chemistry were made by amateurs with sufficient private wealth to buy their own supplies and equipment and spend the time needed to carry on their research. Until the late 20th century, it was rare for governments, universities, or industry to enthusiastically support basic research at the levels it needed to function effectively, one reason that American philanthropy has played such a rare and important role in the advancement of basic research in the United States.

Yet, history has made abundantly clear the undeniable interaction among basic research, applied research, and technology, generating the question as to how these three legs of the scientific enterprise are to be provided for in each generation. This discussion has reached its zenith in the modern world as science and technology have permeated every level of society and become a factor in every major social, political, and economic issue of the day. It is hardly surprising that questions that troubled the ancient Greeks; 16th-century Italians; colonial Americans; and mid-20th-century Russian, German, and British scientists and politicians about the interaction of research and politics continue—and are even amplified—in today's world. In Chapter 2, our discussion will focus on these issues and the ways in which they have arisen in the post–World War II world.

References

Alder, Ken. 2002. *The Measure of All Things: The Seven-year Odyssey and Hidden Error That Transformed the World*. New York: Free Press.

Al-Hassan, Ahmad Husuf, and Donald R. Hill. 1988. *Islamic Technology: An Illustrated History*. Cambridge, UK: Cambridge University Press.

Ancient Greek Technology: Replicas and Models: An Approach to the Marvels of the Ancient Greek Masters. 2000. Thessaloniki, Greece: Technology Museum of Thessaloniki.

Babbage, Charles. 1830. *Reflections on the Decline of Science in England*. Project Gutenberg. http://www.gutenberg.org/ ebooks/1216. Accessed on May 15, 2013.

Bacon, Francis. 1620. *The Great Insaturation*. Internet unpaginated version by the University of Adelaide Library. http:// ebooks.adelaide.edu.au/b/bacon/francis/instauration/. Accessed on May 14, 2013.

Bacon, Sir Francis. 1627. *The New Atlantis*. Internet unpaginated version by Project Gutenberg. http://www.gutenberg .org/files/2434/2434.txt. Accessed on May 20, 2013.

Baker, John A. 1939. "A Counterblast to Bernalism." *New Statesman and Nation*. 18: 174–175.

Bates, Ralph Samuel. 1965. *Scientific Societies in the United States*, 3rd ed. Cambridge, MA: MIT Press.

Bernal, J. D. 1939. *The Social Function of Science*. London: G. Routledge & Sons.

Bernal, J. D. 1958. *World without War*. London: Routledge and Paul.

Bidlack, Richard, and Nikita Lomagin. 2012. *The Leningrad Blockade, 1941–1944: A New Documentary History from the Soviet Archives*. New Haven, CT: Yale University Press.

Bindley, T. Herbert. 1914. *On the Testimony of the Soul and on the "Prescription" of Heretics*. London: Society for Promoting Christian Knowledge. Available online at http://www.tertullian.org/articles/bindley_test/bindley_test_07prae.htm. Accessed on May 12, 2013.

British Association for the Advancement of Science. 1933. *Report of the Annual Meeting, 1933*. http://www.archive.org/stream/reportofbritisha33adva/reportofbrit-isha33adva_djvu.txt. Accessed on May 20, 2013.

Brumbaugh, Robert S. 1966. *Ancient Greek Gadgets and Machines*. New York: Crowell, 1966.

Burns, William E. 2003. *Science in the Enlightenment: An Encyclopedia*. Santa Barbara, CA: ABC-CLIO.

Burns, William E. 2005. *Science and Technology in Colonial America*. Westport, CT: Greenwood Press.

Chafy, Randy. 1997. "Exploring the Intellectual Foundation of Technology Education: From Condorcet to Dewey." *Journal of Technical Education*. 9(1): 6–19. Also available online at http://scholar.lib.vt.edu/ejournals/JTE/v9n1/chafy.html. Accessed on May 15, 2013.

Chatrchyan, S., et al. 2012. "Combined Results of Searches for the Standard Model Higgs Boson in pp Collisions at $\sqrt{s}$ = 7 TeV." *Physics Letters B*. 710(1): 26–48.

"Chronology of Jewish Persecution." Jewish Virtual Library. http://www.jewishvirtuallibrary.org/jsource/Holocaust/Chronology_1932.html. Accessed on May 19, 2013.

Clagett, Marshall. 1971. *Greek Science in Antiquity*. Freeport, NY: Books for Libraries Press (reprint of 1955 edition).

Cohen, I. Bernard. 1956. *Franklin and Newton; an Inquiry into Speculative Newtonian Experimental Science and Franklin's Work in Electricity as an Example Thereof.* Philadelphia: American Philosophical Society.

Condorcet (Marquis de Jean-Antoine-Nicolas de Caritat). 1791. *Cinq mémoires sur l'instruction publique* (*Five Papers on Public Instruction*). http://classiques.uqac.ca/classiques/condorcet/cinq_memoires_instruction/Cinq_memoires_instr_pub.pdf. Accessed on May 15, 2013.

"Create a List." Nobel Prizes. http://www.nobelprize.org/nobel_prizes/lists/all/create.html. Accessed on May 19, 2013.

Dallal, Ahmad S. 2010. *Islam, Science, and the Challenge of History.* New Haven, CT: Yale University Press, 2010.

Derry, T. K., and Trevor I. Williams. 1993. *A Short History of Technology: From the Earliest Times to A.D. 1900.* New York: Dover Publications (reprint of the original 1960 edition).

De Santillana, Giorgio. 1955. *The Crime of Galileo.* Chicago: University of Chicago Press.

Descartes, René. 1637. *Discourse on the Method for Reasoning Well and for Seeking Truth in the Sciences.* Trans. by Ian Johnston. Vancouver Island University. http://records.viu.ca/~johnstoi/descartes/descartes1.htm. Accessed on May 14, 2013.

Dupree, A. Hunter. 1957. *Science in the Federal Government, a History of Policies and Activities to 1940.* Cambridge, MA: Belknap Press. Also available online at http://www.worldcat.org/title/science-in-the-federal-government-a-history-of-policies-and-activities-to-1940/oclc/1084324&referer=brief_results. Accessed on May 16, 2013.

"Early Christian Attitudes to Science." http://www.badnewsaboutchristianity.com/ea0_trad.htm#early. Accessed on May 12, 2013.

Emergency Committee in Aid of Displaced German Scholars. 1936. *List of Displaced German Scholars.* London: Speedee Press Services.

Emergency Committee in Aid of Displaced German Scholars. 1937. *Supplementary List of Displaced German Scholars.* London: Speedee Press Services.

"Emergency Committee in Aid of Displaced Foreign Scholars Records, 1927–1949." New York Public Library. http://www.nypl.org/sites/default/files/archivalcollections/pdf/emergency.pdf. Accessed on May 19, 2013.

Evans, Richard. "A Short History of Technical Education." http://technicaleducationmatters.org/series/a-short-history-of-technical-education/. Accessed on February 14, 2014.

"Examples of Anti-Semitic Legislation, 1933–1939." Holocaust Encyclopedia. http://www.ushmm.org/wlc/en/article.php?ModuleId=10007459. Accessed on May 19, 2013.

Farrington, Benjamin. 1939. *Science and Politics in the Ancient World.* New York: Barnes & Noble, Inc.

Feibleman, James Kern. 1961. "Pure Science, Applied Science, Technology, Engineering: An Attempt at Definitions." *Technology and Culture: The International Quarterly of the Society for the History of Technology.* 2(4): 305–317.

Filner, Robert E. "Science and Marxism in England, 1930–1945." http://www.autodidactproject.org/other/sn-filner.html. Accessed on May 21, 2013.

Finocchiaro, Maurice A., ed. 2008. *The Essential Galileo.* Indianapolis: Hackett Publishing Company.

Freely, John. 2012. *The Flame of Miletus: The Birth of Science in Ancient Greece (And How It Changed the World).* London; New York: I. B. Tauris.

Galilei, Galileo. 1623. *Il Saggiatore [The Assayer].* Rome: Giacomo Mascardi.

Galilei, Galileo. 1638. *Discorsi e dimostrazioni matematiche, intorno à due nuove scienze [Discourses and Mathematical*

Demonstrations Relating to Two New Sciences]. Leiden: Elzevir.

"Germany: Jewish Population in 1933." United States Holocaust Memorial Museum. http://www.ushmm.org/wlc/en/article.php?ModuleId=10005276. Accessed on May 19, 2013.

"Gesta Lynceorum." National Committee for the Fourth Centenary of the Founding of the Academy of Lincei. http://www.lincei-celebrazioni.it/igesta.html. Accessed on May 13, 2013.

Gillispie, Charles Coulston. 2004. *Science and Polity in France: The Revolutionary and Napoleonic Years*. Princeton: Princeton University Press.

Gower, Barry. 2005. *Scientific Method: A Historical and Philosophical Introduction*. New York: Routledge.

Greenberg, Daniel S. 1967. *The Politics of Pure Science*. New York: The New American Library.

Harris, William. "How the Scientific Method Works." How Stuff Works. http://science.howstuffworks.com/innovation/scientific-experiments/scientific-method.htm. Accessed on May 13, 2013.

Hartley, Anthony. 1963. *A State of England*. New York: Harcourt, Brace & World.

Hirsch, Francine. 2005. *Empire of Nations Ethnographic Knowledge & the Making of the Soviet Union*. Ithaca, NY: Cornell University Press.

"History of Russian Academy of Sciences." http://www.npd.ac.ru/npd/history.npd/hisras_e.htm. Accessed on May 18, 2013.

Huxley, Julian. 1934. *If I Were Dictator*. New York; London: Harper & Brothers.

International Human Genome Sequencing Consortium. 2006. "Initial Sequencing and Analysis of the Human Genome." *Nature*. 441(7093), Supp (June 1, 2006): 30–92.

Jones, Richard. "On Pure Science, Applied Science, and Nano-technology." Soft Machines. http://www.softmachines.org/wordpress/?p=926. Accessed on May 15, 2013.

Joravsky, David. 1960. "Soviet Scientists and the Great Break." *Daedalus*. 89(3): 562–580. Adapted from David Joravsky. 1961. *Soviet Marxism and Natural Science, 1917–1932*. New York: Columbia University Press.

Joravsky, David. 1970. *The Lysenko Affair*. Cambridge, MA: Harvard University Press.

Kotsanas, Kostas. "Ancient Greek Technology." http://www.kotsanas.com/. Accessed on May 11, 2013.

"Law for the Restoration of the Professional Civil Service, April 7, 1933." Documents of the Holocaust. http://www.yadvashem.org/about_holocaust/documents/part1/doc10.html. Accessed on May 19, 2013.

Lecourt, Dominique. 1977. *Proletarian Science?: The Case of Lysenko*. London: NLB; Atlantic Highlands, NJ: Humanities Press.

"List of Inventions and Inventors During the Industrial Revolution (1800–1899)." http://www.docstoc.com/docs/105402338/listofinventorsandinventions. Accessed on May 15, 2013.

Mays, Wolfe. 1974. "Scientific Method in Galileo and Bacon." *Indian Philosophical Quarterly*. 1: 217–239. Also available online at http://unipune.ac.in/snc/cssh/ipq/english/IPQ/1-5%20volumes/01-3/1-3-3.pdf. Accessed on May 13, 2013.

McGucken, William. 1978. "On Freedom and Planning in Science: The Society for Freedom in Science, 1940–46." *Minerva*. 16(1): 42–72.

Medawar, J. S., and David Pyke. 2001. *Hitler's Gift: The True Story of the Scientists Expelled by the Nazi Regime*. New York: Arcade Publishers.

Medvedev, Zhores A. 1969. *The Rise and Fall of T. D. Lysenko.* New York: Columbia University Press.

Mendes-Flohr, Paul R., and Jehuda Reinharz. 2011. *The Jew in the Modern World: A Documentary History.* New York: Oxford University Press.

Metzger, Walter P. 1955. *Academic Freedom in the Age of the University.* New York: Columbia University Press.

Micklos, David A. 1988. *The First Hundred Years: A History of Man and Science at Cold Spring Harbor Laboratory.* Cold Spring Harbor, NY: Cold Spring Harbor Laboratory.

Moser, Petra, Alessandra Voena, and Fabian Waldinger. *German-Jewish Émigrés and U.S. Invention.* http://cliometrics.org/assapapers/Moser%20Voena%20Waldinger.pdf. Accessed on May 19, 2013.

Nasr, Seyyed Hossein. 1976. *Islamic Science.* London: World of Islam Festival Publishing Company.

Polanyi, Michael. 1962. *The Republic of Science, Its Political and Economic Theory.* Chicago: Roosevelt University.

Price, Don K. 1954. *Government and Science: Their Dynamic Relation in American Democracy.* New York: New York University Press.

Rayward, W. Boyd. 1999. "H.G. Wells's Idea of a World Brain: A Critical Reassessment." *Journal of the American Society for Information Science.* 50(7): 557–573. Available online at http://people.lis.illinois.edu/~wrayward/Wellss_Idea_of_World_Brain.htm. Accessed on May 21, 2013.

Reingold, Nathan. 2008. "Joseph Henry." In Charles Coulston Gillispie, Frederic Lawrence Holems, and Noretta Koertge, eds. *Complete Dictionary of Scientific Biography.* Detroit, MI: Charles Scribner's Sons. Available online at http://www.encyclopedia.com/topic/Joseph_Henry.aspx. Accessed on May 16, 2013.

Reinisch, Jessica. "The Society for Freedom in Science, 1940–1965." http://www.academia.edu/1088487/The_Society_for_Freedom_in_Science_1940-1965. Accessed on May 21, 2013.

"The Richest Americans." CNN Money. http://money.cnn.com/galleries/2007/fortune/0702/gallery.richestamericans.fortune/index.html. Accessed on May 16, 2013.

Rossi, Paolo. 1957, 1968. *Francis Bacon: From Magic to Science*. Chicago: University of Chicago Press.

Saint Augustine. *Handbook on Faith, Hope, and Love*. Trans. by Albert C. Outler. Grand Rapids, MI: Christian Classics Ethereal Library. Available online at http://www.ccel.org/ccel/augustine/enchiridion.pdf. Accessed on May 12, 2013.

Saliba, George. 2007. *Islamic Science and the Making of the European Renaissance*. Cambridge, MA: MIT Press, 2007.

Schofield, Malcolm. 2007. "The Noble Lie." In G. R. F. Ferrari, ed. *The Cambridge Companion to Plato's* Republic. Berkeley: University of California, Chapter 6.

Schofield, Robert E. 1957. "The Industrial Orientation of Science in the Lunar Society of Birmingham." *Isis*. 48(4): 408–415.

Selin, Helaine, ed. 2000. *Astronomy across Cultures: The History of Non-Western Astronomy*. Dordrecht; Boston: Kluwer Academic Publishers, 2000.

Sheehan, Helena. 1985. *Marxism and the Philosophy of Science: A Critical History*. Atlantic Highlands, NJ: Humanities Press.

Soyfer, Valery. 1994. *Lysenko and the Tragedy of Soviet Science*. New Brunswick, NJ: Rutgers University Press.

Thurgood, Lori, Mary J. Golladay, and Sustan T. Hill. 2006. *U.S. Doctorates in the 20th Century*. Arlington, VA: National Science Foundation, Division of Science Resources Statistics.

Tibbetts, Gary G. 2013. *How the Great Scientists Reasoned: The Scientific Method in Action.* Amsterdam; Boston: Elsevier.

Year Book—Carnegie Institution of Washington. 1904. Washington, DC: Carnegie Institution of Washington, vol. 3.

One might say, for instance: for "Holy Office" read "AEC Board,"
for "Caccini" read "Crouch," for "Lorini" read "Borden," for "S.J."
(Societas Jesu [Society of Jesus, or Jesuits]) read "SAC" (Strategic Air
Command), for "Report of the Preliminary Commission" read "Gray-
Morgan Majority Report," for "Grienberger" read "Teller," for "cer-
tain German mathematicians" read "Dr. Malraux," and so on.

(De Santillana 1955, viii)

This passage is taken from the introduction to Giorgio de Santillana's brilliant book, *The Crime of Galileo*, in which the author shows the close connection between two trials of scientists
separated by a space of more than five centuries, the trials of
Galileo and J. Robert Oppenheimer. The latter case had been
decided only a year prior to publication of de Santillana's book,
and it could not have provided a more clear example of the fact
that nothing is new in the world. The attack by politicians on
Galileo in the mid-16th century had been played out in remarkable similarity by politicians on Oppenheimer in the early
1950s. The "Caccini" mentioned by de Santillana, for example,
was a Dominican monk, Father Tommaso Caccini, who had already been disciplined as a "scandal-maker" by the Archbishop
of Bologna before he decided to start accusing Galileo of heresy

---

General Leslie Groves (r) and Dr. J. R. Oppenheimer examine the remains of the tower from which a test atomic bomb was detonated near
Alamogordo, New Mexico, in 1945. (Library of Congress)

in 1614 (de Santillana 1955, 42), while "Crouch" was Paul Crouch, an organizer for the Communist Party in California and one of the first individuals to accuse Oppenheimer of possibly being an agent for the Communist Party. (Both men were eventually revealed to be more interested in causing trouble than providing factual information about the cases involved.) (Alsop and Alsop 1954) Similarly, the "Holy Office" mentioned here is the ecclesiastical court of the Roman Catholic Church, before which Galileo appeared, and by whom he was convicted, and the "AEC Board" was Personnel Security Board that removed Oppenheimer's security clearance for being an untrustworthy employee. And the two "reports" were the final judgments of the special committee created to advise the inquisition on the Galileo matter in 1616, on the one hand, and the majority decision of the Personnel Security Board in 1954, on the other.

Galileo Revisited: The Case of J. Robert Oppenheimer

The Science behind the Politics

The Oppenheimer Case, as we will call it, had its origins in the 1930s, during a period when physicists and chemists were fascinated by a newly discovered technology called the n,γ reaction. In such reactions, a neutron, one of the basic particles of which atoms are made, is fired at the nucleus of an element, such as the nucleus of a copper atom, and a gamma ray (γ) is given off. In the event, the target element (copper in this example) changes into a radioactive form which, in turn, often gives off other particles and itself changes into a new element one place higher in the periodic table. Scientists were excited by this reaction because it gave them a relatively simple way of converting one element into another element, a goal of the ancient science of alchemy dating back hundreds of years. During the 1930s, then, researchers were bombarding just about every element in the periodic table with neutrons to see what results they could observe.

The one n,γ reaction of special interest to many researchers involved the bombardment of the element uranium with neutrons. If that experiment followed the pattern of other such reactions, the new element produced would be one unit heavier than the heaviest known natural element, uranium. That is, successful conduct of the experiment would have resulted in the formation of a new, artificial element, . . . which is just the result that Italian physicist Enrico Fermi thought he had obtained in a uranium-n,γ experiment in 1934. He turned out to be wrong, but three German colleagues, Lise Meitner, Otto Hahn, and Fritz Strassman, obtained similar results four years later. Meitner, Hahn, and Strassman were fairly certain that they had not observed a traditional n,γ experiment, but were unwilling to accept what seemed to be the correct explanation: that the uranium nucleus had not just given off a gamma ray, but had actually split into two large parts, a barium atom and a krypton atom. No one had ever observed the *fissioning* of an atom like this. In fact, it was not until Meitner and her nephew Otto Frisch (both Jews who had left Germany to escape Nazi persecution) developed a mathematical theory to explain the event did their colleagues recognize that such an event, the fissioning of an atom, had actually occurred.

This discovery came at an extraordinary moment in human history with potentially profound consequences for the future of the world. Virtually all of the basic research on nuclear fission was taking place in Germany and Italy by scientists who fairly quickly realized the potential application of the process. Rudimentary calculations show that the fissioning of an atom like uranium is interesting and important not only for theoretical reasons, but also because the reaction is always accompanied by the release of quantities of energy unmatched by any other process known to humans. It takes little effort for someone with the right mind-set to realize at least one application of the reaction: the construction of a weapon, a weapon that would be more powerful than anything humans had ever imagined. (As an example, the first fission bomb ever used in warfare, dropped on Hiroshima,

Japan, in August 1945, had a power that was measured in kilotons, thousands of tons of TNT equivalence. That bomb was a 20 kiloton bomb, meaning it had the power equivalence of 20,000 tons of TNT.)

The discoveries associated with nuclear fission were taking place in Germany and Italy, then, precisely at the time when new governments in both nations had just taken over and were planning an attack on the rest of Europe with the goal of subjugating those nations and creating a new continent under Fascist dominance. Gaining a weapon such as the fission bomb would provide the means by which that goal could be achieved with relatively modest effort (but unimaginable destruction and death).

Only one problem stood in the way of this scenario, one that was not fully understood until the end of World War II in the late 1940s: Adolf Hitler was suspicious of the "Jewish science" on which the theory of nuclear fission was based, and he did not trust the investment of time, money, and personnel in the development of a weapon based on that kind of science. As a result, development of a fission bomb never progressed very rapidly in Nazi Germany and, in any case, not rapidly enough to produce a working bomb before the war ended.

Such was not the case in many other parts of the world, however. Because of its inherently international nature (scientists from all nations communicate regularly and openly with each other), physicists and chemists in France, England, the United States, and other parts of the world knew about the Hahn/Strassman/Meitner/Frisch discovery, recognized the potential advantage it gave to Nazi Germany, and began to panic about the possible consequences of the new research for the free world.

Reaction in the United States

For many scientists who knew about the process of nuclear fission and understood its potential for use in building a weapon, the United States seemed to be the logical first line of defense against the development and use of a nuclear weapon. The problem was how to convince American authorities of the threat this scientific breakthrough represented to the free world.

The leaders in this effort to get the United States moving on a fission project were Italian-born physicist Enrico Fermi and Hungarian-born physicist Leó Szilárd, both then on the faculty at Columbia University, in New York City. In one of the first efforts to interest American officials in this issue, Fermi met with Admiral S. C. Hooper, Director of the Technical Division of Naval Operations on March 17, 1939. He presented the scientific information about nuclear fission and explained its potential in the construction of a weapon. Hooper showed no interest in the issue, and there was no follow-up on Fermi's efforts (Jungk 1958, 78).

This failure soon took on more significance when Fermi, Szilard, and their colleagues received information only a month later that a committee of scientists in Germany had been appointed to begin studying just the potential project they had feared: the use of nuclear fission to develop a weapon. They had no way of knowing that this effort had no government support and would lead to no significant research on the topic. They knew only enough to increase their concerns about the development of a fission weapon by Germany.

By August of 1939, Szilard, Fermi, and their colleagues had come up with a plan for forcing their concerns on the U.S. government. They (primarily Szilard) decided to write a letter to President Franklin D. Roosevelt describing nuclear fission and its potential use in constructing a weapon. They decided to get the most famous scientist of the era, Albert Einstein, to sign the letter, which was delivered to the president's office on August 2, 1939. Roosevelt did not actually see the letter for more than two months, however, because of the outbreak of World War II on September 1. He then appointed a special committee to study the problem, which found $6,000 to fund research on fission by a team led by Fermi, but went no further on the project.

Not much of anything happened on a possible bomb project, in fact, for more than two years. On January 19, 1942, Roosevelt finally authorized the initiation of a project to construct a weapon based on nuclear fission. That project, given the name of the *Manhattan Project*, was under the control of U.S. Army and supervised by Brigadier General Leslie R. Groves, Jr., who, in turn, appointed physicist J. Robert Oppenheimer to lead all scientific aspects of the project.

Oppenheimer, the Manhattan Project, and Reaction

At the time, Oppenheimer was one of the most highly respected theoretical physicists in the world, having already made a number of contributions in the fields of quantum mechanics, wave mechanics, and nuclear fission. He had very little experience in administration, but quickly mastered the techniques required to obtain the maximum output from the hundreds of physicists, engineers, and technicians placed under his authority. Less than three years after assuming his position in the project, Oppenheimer had led his team of researchers to the first test of a nuclear weapon, the so-called Trinity nuclear test, held on July 16, 1945. The test was a complete success, and the two remaining bombs that had been constructed by that time (known as *Little Boy* and *Fat Man*) were ready for combat use. The first of these bombs was dropped on Hiroshima, Japan, on August 6, 1945, resulting in an estimated 66,000 deaths and 69,000 injuries and untold physical damage to the city. Three days later, Fat Man was dropped on Nagasaki, Japan, producing an estimated 39,000 deaths and 25,000 injuries and, again, massive physical damage to the city. (Considerable disagreement exists about the number of dead and wounded as a result of the bombings. See, for example, Ford 2013. For a much more complete telling of the story of the Manhattan Project and its follow-up, see Jungk 1958.)

In one respect, the bombing of Hiroshima and Nagasaki brought an end to the story that started with Albert Einstein's letter to President Roosevelt. American scientists (aided to a very large extent by emigres from European countries—especially Germany and Italy) brought the theoretical concept of nuclear fission into reality, producing a weapon that essentially brought World War II to an end. From another standpoint, the successful completion of the Manhattan Project was only the beginning of a whole new relationship between scientists, politicians, the government, and the general public. One of the most important elements in that change was the sudden realization by scientists of the moral and political consequences of the research they had just completed. Oppenheimer himself has often been quoted about the Trinity test. He said that, watching the mushroom cloud forming over the New Mexico desert, he thought of a line from the Indian classic, the *Bhagavad Gita*, in which the god Vishnu says, "Now I am become death; the destroyer of worlds." Oppenheimer went on to say, "I suppose we all thought that, one way or another" ("Now I am become death . . ." 2013).

The post–World War II world was a very different place from the one in which the Manhattan Project had achieved such astounding success, and Oppenheimer found his status in that new political climate also drastically changed. That fact became abundantly clear on December 23, 1953, when Lewis L. Strauss, chairman of the Atomic Energy Commission (AEC), informed Oppenheimer that his security clearance was being suspended pending resolution of issues related to his potential risk in working in the United States government. The government had decided that Oppenheimer's Communist's leanings posed a threat to national security. Strauss explained that Oppenheimer's case would be heard before the AEC's Personnel Security Board (PSB) on charges that he was guilty of treason against the United States. (A number of excellent books and articles are available on the details of the Oppenheimer case. See, for example, Bird and Sherwin 2005; Carson and Hollinger 2005; Jungk 1958; Polenberg 2002; and Stern 1969. A complete transcript of the case is available online at United States Atomic Energy Commission 1954.)

The hearing turned out to be a farce, not so very different from the so-called show trials held in the Soviet Union in the late 1930s, in which defendants were essentially found guilty, with or without evidence, generally prior to the trial itself, at least in the minds of the controlling authority (Wales 2013). In Oppenheimer's case, the federal government had been collecting evidence about his "disloyalty" for at least a decade. In his earlier years, Oppenheimer, like many intellectuals, had become enamored of the Soviet system and a supporter of and contributor to a variety of Communist causes. Both his brother and his wife had been members of the Communist Party ("the Party").

During the pre–World War II years, Oppenheimer himself had been involved in some complex and rather unsavory associations with members and "sympathizers" of the Party and had even revealed the name of one such person (erroneously) to the U.S. government. Also, like many intellectuals, however, he had become disillusioned by the Soviet–Nazi nonaggression pact of 1939 and discontinued his support for the Communist Party.

Oppenheimer's involvement with the Party was well known to the U.S. Federal Bureau of Investigation (FBI) which collected a massive dossier on his activities over a period of at least 11 years and consisting of more than 10,000 pages (Mckie 2013; Stern 1969, 2). During the war, however, none of this information was considered serious enough for governmental action. General Groves highlighted at least one reason for this inaction. "He is," Groves said, "absolutely essential to the [Manhattan] project" (Kevles 2013, quoting Herken 2002). Probably more to the point, almost none of Oppenheimer's colleagues or associates had the slightest doubt about Oppenheimer's loyalty to the United States (Jungk 1958, 321).

The end of the war saw a change in the nation's political tenor, however. This change was reflected in the Republic Party's national nominating convention of 1952, at which the party promised to conduct a massive sweep of the federal government, exposing and removing every unreliable "subversive" in the U.S. government. When Dwight D. Eisenhower was elected President, Republicans began to carry out that promise, and Oppenheimer became a prime object of their efforts. At the PSB hearings, years of FBI records of Oppenheimer's pro-Communist activities, along with a collection of misinformation and outright lies, were presented to prove his disloyalty. His case was not helped by the active opposition of one of Oppenheimer's most famous colleagues (and subordinates in the Manhattan Project), Hungarian-born Edward Teller. Known as the Father of the Hydrogen (fusion) Bomb, Teller had long been frustrated by Oppenheimer's opposition to research on fusion weapons, and told the PSB that his opposition was evidence of his former boss's "unreliability" (Wales 2013).

The investigating committee's decision was largely a foregone conclusion, and on May 27, 1954, it voted two-to-one, to recommend to the AEC that Oppenheimer's security clearance be withdrawn. It found that 20 of the 24 charges against him were either "true" or "substantially true" ("Findings and Recommendations of the Personnel Security Board in the Matter of Dr. J. Robert Oppenheimer" 2013).

The debate over the Oppenheimer case has never really come to a conclusion, with some observers still arguing that he posed a real and serious threat to the security of the United States (Burress 2013). The majority of historians, however, probably feel that the PSB decision was largely a political circus that resulted in a gross injustice to the man who has become known as the Father of the Atomic Bomb. Perhaps more to the point for the Oppenheimer case is what it has to say about the relationship of science and politics in the United States. As one observer, T. C. Wales, has written, "Any security regime that stifles the free exchange of ideas, throws a blanket of suspicion over an entire class of people, or becomes subject to political manipulation, will constitute a far greater threat to the American experiment than the enemies it is designed to frustrate" (Wales 2013).

Nor is the Oppenheimer case purely an interesting relic of U.S. history. As Wales continues in his analysis of the case, "From the failed prosecution of the Los Alamos scientist Wen Ho Lee in 1999 for stealing nuclear secrets, to the continuing persecution of Professor Sami Al-Arian for allegedly associating with Palestinian terrorists, Americans continue to mistake innuendo for guilt" (Wales 2013; for more on the two cases cited, see Fisher 2013 and Lee and Zia 2001).

The Oppenheimer case (and the trial of Galileo, nearly five centuries earlier) illustrates how governmental officials and politicians can silence (or, in cases like that of Trofim Lysenko, promote) a powerful scientist, regardless of the contributions he or she has made (or, in the case of Lysenko, not made) in the field of science. It made no difference that Oppenheimer had led one of the greatest scientific projects in human history, bringing victory to the United States in World War II, or that he was one of the leading theoretical physicists of his time. What mattered to his accusers was that his political views (or at least, what they thought those views were) were unacceptable to leading figures in government.

The Space Race and the Apollo Program

But the silencing of "wrong thinking scientists" is only one of the many ways in which politics and science can interact. Consider now a second possibility that follows.

On May 25, 1961, President John F. Kennedy sent a message to the U.S. Congress dealing with a number of "urgent national needs." Among the urgent needs the president identified was a new space program designed to place a human on the Moon before the end of the decade (before 1970). He said that such a program "will be more impressive to mankind, or more important for the long-range exploration of space; and none will be so difficult or expensive to accomplish" ("Special Message to the Congress on Urgent National Needs" 2013).

In one fell swoop, the president laid before Congress a program that was to become the largest, most expensive, most labor-intensive peacetime science project in the history of the world. In essence, he was describing a research project that would consume the energy of thousands of scientists for more than a decade. The interesting point about Kennedy's speech was that he had very little record of any interest in scientific research prior to his becoming president, and certainly little to say about the role scientific research might have in the American economy. Where did this grandiose notion, which was to overwhelm a large part of the scientific community for such a long period of time, come from?

The point to be made at the outset was that the Moon project (later to take the form of the Apollo Space Program) was never intended to be primarily a *scientific* program; it was, instead, a component of the president's political agenda for his first term in office.

During his campaign for reelection to the U.S. Senate from Massachusetts in 1958, Kennedy had seized upon an issue that he hoped would win him votes, the "missile gap" that President Dwight D. Eisenhower had allowed to develop between the United States and the Soviet Union. The term *missile gap* was apparently first used by Kennedy in a speech on August 14, 1958, when he used information collected by the U.S. military predicting that the Soviet Union would have the ability to produce as many as 500 intercontinental ballistic missiles (ICBMs) by 1961 or 1962, while the United States would have fewer than a hundred. Columnist Joseph Alsop provided even more frightening data, predicting that the Soviet Union might have as many as 1,500 missiles within a few years, with the United States having only 130 (Alsop 1959). As it turned out, Alsop (and a number of his colleagues) was dramatically wrong: At the time of his prognostication, the Soviet Union had only four operational ICBMs, about 2 percent of the number Alsop and other "experts" were predicting (Day 2013).

Regardless of the actual data, Kennedy continued to repeat his erroneous statements about the "missile gap," both in his senatorial campaign, and in his presidential campaign against Eisenhower's vice president and Republican presidential candidate in 1961. Given Kennedy's extraordinarily small margin of victory in that campaign (118,550 out of more than 68 million popular votes cast), it is probably fair to say that the missile gap argument may have been a determining factor in the election.

But, once elected, Kennedy found that he could not do anything about the missile gap, because it didn't exist. He also faced another, perhaps more serious, dilemma. By the time Kennedy was sworn into office on January 20, 1951, the Soviet Union had made some very real steps forward in space science. The American public, which had been used to seeing the United States as being first in just about every field of science since the end of World War II, was faced with the fact that the Soviet Union had accumulated a number of important "firsts" in space science, including:

> August 21, 1957: Launch of the first ICBM, capable of carrying weapons thousands of miles around the world or objects and humans into near space
>
> October 4, 1957: Launch of the first artificial satellite, Sputnik
>
> November 3, 1957: Launch of the first animal into space, the dog Laika
>
> January 2, 1959: Launch of the first lunar spacecraft, Luna 1
>
> September 14, 1959: Delivery of the first manmade object to the surface of the Moon, Luna 2
>
> April 12, 1961: First human in space, Yuri Gargarin, in Vostok 1

The challenge Kennedy faced, then, was to find a project through which the United States could reassert its dominance in science in general, and space science in particular. The goal of the project he decided on was to put a human on the Moon before the end of the decade. The scientific objectives of that project were very much less clear, and for Kennedy, they were largely irrelevant. The Apollo program was conceived of, and remained throughout its existence, primarily a political project, not a scientific program ("Project Apollo: A Retrospective Analysis" 2013).

Virtually any student of history recognizes that Kennedy was correct in assessing the popular response to his proposal. The Congress, the scientific community, and the general public were all, for the most part, excited by the prospects of this potential breakthrough in space science. But the project also had its critics from all three areas. One of the most common complaints was that the United States had more than enough practical problems to deal with (hunger, poverty, infrastructure, education, health, and so on) as it was. The billions of dollars planned for the space program could better be spent, the argument went, on domestic issues such as these.

For many scientists, however, the most troubling issue was one that had been around since the birth of modern science: Who is it that should decide what research scientists should work on? As we have seen in Chapter 1 of this book, many nations have decided that the answer to that question is: government. As Stalin, Hitler, and a number of governmental leaders, as well as many leading scientists themselves, have concluded, the primary goal of science is to serve the needs of society. Prior to Kennedy's speech, this principle meant having scientists work on more abundant supplies of food, better means of preventing and treating disease, improved methods of transportation and communication, better domestic products for consumers, and the like. But there had been few, if any, situations in which scientists were marshaled to work on research primarily to improve a nation's prestige.

In addition to this philosophical issue, scientists were worried about the ways in which Apollo might change the balance of research funding in the United States. If the Congress was going to be asked to spend a few hundred millions or billions of dollars a year for nearly a decade to put one person on the Moon, how would such an effort affect funding for other types of space research, or, for that matter, other types of scientific research in general? No one knew the answer to that question, because such issues had not been a factor in making the decision to send a person to the Moon. (For more on this issue, see "America Starts for the Moon: 1957–1963" 2013.)

In some respects, the Apollo program really ended on July 21, 1969, when astronauts Neil Armstrong and Buzz Aldrin stepped on the surface of the Moon. With that act, President Kennedy's mission had been completed, a fact confirmed when Mission Control personnel in Houston flashed the sign TASK ACCOMPLISHED, July 1969, when the astronaut's arrival had been confirmed ("Project Apollo: A Retrospective Analysis" 2013). Additional flights continued, of course, but after Apollo 11, the U.S. Congress and the general public began to lose interest and funding for additional flights became ever more difficult to defend and obtain. In fact, the last three flights planned for the program, Apollo 18, 19, and 20, were cancelled for lack of funds.

Big Science, Little Science

Both the World War II Manhattan Project and the postwar Apollo Space Program marked the beginning of a new era in scientific research. It was an era that was first called *Big Science* in a 1961 article by then director of the Oak Ridge National Laboratory, Alvin Weinberg. Weinberg noted that the scope of scientific research during and just following the war had not just increased incrementally, but had exploded. The total cost of the Apollo program, for example, was estimated at $19,408,134,000, accounting for just over a third of the total NASA budget during the program's lifetime ("The Apollo Program" 2013). At its peak level of activity, more than 400,000 people were involved in the Apollo program in one capacity or another, and more than 20,000 institutions of higher education and industrial firms were making contributions to the project ("NASA Langley Research Center's Contributions to the Apollo Program" 2013).

Big Science was (and is) distinguishable, then, by the large number of individuals and research organizations involved in a project, the size of the project's budget, the size of the machinery used to conduct research, and the size of the physical space required to carry out the research. Big Science, thus, differed very significantly from Little Science, the type of science that was characteristic of the pre–World War II period, in which an individual researcher or handful of researchers worked in a small university or industrial laboratory on a project that might cost no more than a few tens or hundreds of thousands of dollars. (The classic work on the transition from Little Science to Big Science is Derek de Solla Price's *Little Science, Big Science* 1963.)

Big Science differed from Little Science in another important way also. With so much money involved and so many scientists and technicians employed, Big Science projects stood a much greater chance of impacting the social, political, and economic system of a nation. An interesting example of that truism began to unfold just as the Apollo program was coming to an end: the planned construction of the Superconducting Super Collider (SSC).

The Superconducting Super Collider

The Superconducting Super Collider (SSC) was to be a machine designed to study the basic nature of matter. It was a type of particle accelerator similar to linear accelerators, cyclotrons, betatrons, synchrotrons, and other devices capable of propelling electrons, protons, and other small particles to speeds close to the speed of light. When such rapidly moving particles collide with each other or with stationary targets, they tear atoms apart into their basic constituents. Such experiments provide scientists with information as to the most fundamental composition of matter, information that goes well beyond the protons, neutrons, and electrons in atoms with which most people are familiar.

It is the nature of particle physics (the field of science devoted to the very smallest particles of matter) to require ever larger and larger machines to gain new knowledge. By the early 1980s, particle physicists had come to the conclusion that none of the existing particle accelerators available in the United States—or, indeed, the world—was powerful enough to go one level deeper into the structure of the atom. No new knowledge could be obtained from existing technology. As a result, these researchers proposed the construction of an even larger and more powerful particle accelerator, the SSC.

The proposed new accelerator would consist of a ring nearly 83 km (52 miles) in diameter consisting of a concrete tunnel into which was built the accelerating ring itself. (By comparison, the largest existing accelerator, the Tevatron at the Fermilab outside Chicago, had a ring 6.4 km [4.0 mi] in diameter.) The ring was to consist of two metal tubes a few centimeters in diameter and 70 cm (28 in.) apart, through which particles were to move. Those particles were to be accelerated to energies of at least 20 TeV (teraelectron volts; trillion electron volts). By comparison, the maximum energy obtainable from the Tevatron was about 1 TeV. With particles of this energy, physicists hoped and expected to discover a number of even more basic particles than those already known, especially the prize discovery of the Higgs boson, expected to be one of the most fundamental particle discoveries ever made (and one that was finally announced in March 2013 at the Large Hadron Collider in central Europe). (Probably the best single overview of the SSC still available is Perl 1968.)

The total cost of the SSC was originally to be about \$4.375 billion (Boesman 1987, CRS-7). What arguments might one make to the U.S. Congress to convince it to allocate that amount of money for the project? The most honest argument, of course, would have been simply that particle physicists badly needed the new machine to obtain breakthroughs in their understandings of the composition of matter, an argument about basic research. But such arguments were often difficult to make before legislators who also had to deal with the practical issues in their constituents' lives, such as health, education, jobs, and the economy. At one point in the debate, for example, Senator Dale Bumpers (D-Ark.) said that "It would be nice to know the origin of matter. It would even be nicer to have a balanced budget" (Rubin and Idelson 1991, 1893).

Instead, proponents of the project tended to emphasize the practical benefits of the project, even though that argument was sometimes a stretch. In one press release, for example, the Department of Energy (DOE) noted that "past investments in studies of the interior of atoms have been repaid hundreds of times over in terms of new knowledge, new technologies, new jobs, national security, advances in medicine, and financial returns to the Treasury" (quoted in Bazell 1987, 10). Supporters of the project cited such advances as medical diagnostic techniques, cancer therapy, superconducting cable technology, very large scale integrated circuits, new high tech materials, and a host of applications in environmental and computer sciences (by the Cato Foundation, cited on "Superconducting Super Collider" 2013).

The argument was also made that construction and operation of the SSC would be a significant source of new jobs for skilled and semiskilled personnel, an argument that convinced enough state governments to engender a vigorous bidding war for siting the machine in their region. Finally, proponents of the project appealed to national pride, arguing that the nation's leading physicists would leave the country if the most up-to-date technology was not available for their research. For the most part, the most muted of the benefits and the real reason for which the machine was being built—to advance the field of particle physics—was rarely emphasized to decision makers. In fact, Alvin W. Trivelpiece, Director of the Office of Energy Research at the DOE, is said to have declined to put forward that argument because "it would have been undignified" (Marbach 1987, 45).

Of course, the SSC also had its opponents. As one might expect, one common argument was that the nation should not be spending billions of dollars on pure research projects when basic needs of its citizens were still not being met. This argument gained strength in 1990 when the U.S. Congress passed the Omnibus Budget Reconciliation Act, which imposed severe spending restrictions on major parts of the U.S. economy, including defense spending, nondiscretionary spending (such as Social Security and Medicare), and research and training program. At the same time, however, Congress was still increasing its financial commitments to the construction of the SSC.

Interestingly enough, some of the strongest opposition to the SSC project came from scientists themselves, who saw the program as taking funds from other fields of science to pay for a single enormous particle physics project. Perhaps the strongest statement in this regard came from materials scientist Rustum Roy, then at Pennsylvania State University. At one point, Roy said that he thought the SSC was "a supertoy for a tiny fraction of the spoiled brats of the engineering and science community" (Stipp 1988, 1).

This opposition was not strong enough to prevent work on the SSC from getting underway on a 6,500 hectare (16,000 acre) plot of land near Waxahachie, Texas, in 1991. Only two years later, estimates of the total cost of the project had nearly tripled, reaching about $12 billion. Nearly half of the original estimate, $2 billion, had already been spent in the early stages of construction. The U.S. Congress had had its fill of particle physics for the time and on October 21, 1993, it voted to discontinue funding for the project. The SSC was dead (Mittelstadt 1993, 7; for one of many analyses on the causes for the cancellation of the SSC, see Willard 1994). The land on which the accelerator was to have been built sits unoccupied and undeveloped except for a 135-acre section sold in 2012 to the Magnablend Corporation, which intended to move its facilities to the site (Horning 2012).

The Strategic Defense Initiative

The Manhattan Project and the effort to build the SSC were two of the first really big Big Science projects in the United

States (ultimately to be followed by similar projects, such as the war on cancer, the Human Gene Project, and so on). Almost all of these projects involved debates over how best to spend federal money and human resources, but many also involved questions as to what science had to say about an issue and how best to make use of that scientific knowledge in planning a project. One example that developed close on the heels of the SSC proposal was President Ronald Reagan's Strategic Defense Initiative (SDI), often referred to as his Star Wars program.

For almost four decades after the end of World War II, the primary focus of U.S. foreign policy was the so-called Cold War with the Soviet Union. And an essential feature of that Cold War was the potential use of nuclear weapons by one side or another in an actual shooting war. By the time the first fusion (hydrogen) bombs were being tested in the 1950s, it was apparent to everyone that any future war would be devastating to both sides. Politicians and statesmen from every nation looked for ways to avoid such a war. At first, the only answer to such a dilemma appeared to be an event that came to be known as *mutually assured destruction* (MAD), an occurrence in which, once hostilities got underway, both (or all) combatants were doomed to destruction by the use of nuclear weapons. It became the policy of the United States to construct such an overwhelming arsenal of nuclear weapons to convince the Soviet Union that even if they "won" a war against the United States, it would itself be destroyed in the process.

At the same time, U.S. strategic planners began to think about ways of developing a "shield" that could protect the United States from incoming missiles launched from the Soviet Union. Over the years, various presidents authorized a variety of such programs, including the Nike-Zeus system, under Dwight D. Eisenhower; the Nike-X system, under John F. Kennedy; the Sentinel system, under Lyndon Johnson; and the Safeguard system, under Richard M. Nixon. The challenge of developing a system in which U.S.-launched interceptors could find, attack, and destroy every missile headed for the United States provided to be too much for all of these programs, however, and none ever came close to achieving the objective for which they were created, an all-inclusive protective umbrella covering the nation. (The United States continues to conduct research on such a system under the Missile Defense division of the U.S. National Guard.) ("National Missile Defense" 2013)

When Ronald Reagan came to office as the 40th president of the United States, one of the issues with which he had to deal was the American public's growing uneasiness and discontent with the nation's nuclear military policies. The threat of MAD was, by the time, well known, and efforts to keep the U.S. military system sufficiently in advance of the Soviet Union was causing a severe financial strain on the nation. Early in his administration, Reagan decided that ending the nuclear stalemate with the Soviet Union would be one of his primary objectives as president. He said on a number of occasions that one of the great dreams of his life was to create a nuclear-free war, and one step in that process was to be the development of the evasive missile umbrella that would protect the United States from Soviet missile attacks. At one point, for example, he was quoted as saying that

> Since I knew it would be a long and difficult task to rid the world of nuclear weapons, I had this second dream: the creation of a defense against nuclear missiles, so we could change from a policy of assured destruction to one of assured survival. (Reagan 1990, 550)

The problem was one, of course, that had plagued his predecessors in the office: Was such a program technically feasible? This question was somewhat tainted at the outset because Reagan was not known as someone who held science in the highest regard. His record contains a number of comments that suggest either ignorance of basic scientific information or a tendency to misrepresent that information. Some examples include the following:

- The American Petroleum Institute filed suit against the EPA [and] charged that the agency was suppressing a scientific study for fear it might be misinterpreted. . . . The suppressed study reveals that 80 percent of air pollution comes not from chimneys and auto exhaust pipes, but from plants and trees. (1979; as Presidential candidate)
- Trees cause more pollution than automobiles do. (1981)
- I have flown twice over Mt St. Helens out on our west coast. I'm not a scientist and I don't know the figures, but I have a suspicion that that one little mountain has probably released more sulfur dioxide into the atmosphere of the world than has been released in the last ten years of automobile driving or things of that kind that people are so concerned about. (1980)
- Facts are stupid things. (1988) (Reagan 2013)

In any case, Reagan's eight-year term in office became an ongoing battle between his "dream" of a nuclear defense shield and a significant portion of the scientific community, which continually argued that such a dream was, in fact, truly that, a dream and not a realistic possibility on which federal funds should be spent.

In a speech to the nation on March 23, 1983, Regan outlined his ideas for an SDI, a program for protecting the United States with an array of interceptor rocket that could find and destroy incoming missiles. He acknowledged that the program would be a "formidable task" that might not be achieved before the end of the century, but he assured his listeners that the project was worth "every investment necessary to free the world from the threat of nuclear war" ([Reagan, Ronald] 2013).

Reagan's speech was met with general approval by legislators and the general public, but members of the Congress needed more information about the feasibility and cost of such a system before beginning to appropriate research funds for it. One of their first steps was to ask the Office of Technology Assessment (OTA) to study and report on Regan's proposal. The OTA was created by an act of Congress in 1972 to assist the U.S. Congress in understanding and evaluating science-based projects about which it otherwise had relatively little information. During its 23-year existence, it prepared reports on more than 750 such subjects, providing the technical background on which Congress could make informed decisions (Bimber 1996).

As was its custom, the OTA called on a panel of experts in the field to study the SDI proposal. That panel issued its report in September 1985. Among its major conclusions was that "Assured survival of the U.S. population appears impossible to achieve if the Soviets are determined to deny it to us" (U.S. Congress, Office of Technology Assessment 1985, 33). The fundamental argument was one that would reappear over and over again, namely that it was essentially impossible to develop any type of defense system that would be effective against every missile launched. If even 1 percent of all such missiles were to escape detection and destruction, the resulting devastation would be enormous. The report also pointed to a number of other technical problems, such as the fact that there was really no way to test such a system. One might test the effectiveness of a defensive missile destroying one foreign incoming missile, or perhaps a handful of such missiles, but there would be no way of testing the effectiveness of a system against a full-scale launch like the one of which the Soviet Union was supposedly capable.

The OTA report did not discourage Reagan, a handful of his closest science advisors, or the U.S. Congress, however, and the SDI program was funded in the amount of about $1.4 billion in 1985, $2.8 billion in 1986, $4.8 billion in 1987, and $5.5 billion in 1988 (Pianta 1988, Table 4.6). Nonetheless, members of the scientific community continued to raise objections about the SDI project, based on the waste of money they imagined in pursuing a project with virtually no chance of succeeding technologically. (Within a short period of time, members of the press and the general public began referring to the SDI project as the *Star Wars* project, suggesting a similarity with the science fiction movies of that name, the first three of which had appeared in theaters in 1977, 1980, and 1983.)

The next report on the program by a disinterested scientific group was produced by a special committee of the American Physical Society (APS), which studied the technical details associated with the project. In its 168-page report, the committee was not able to decide one way or another about the feasibility of the project. The science involved was so complex, it reported, that, even under the most favorable circumstances, it would take at least a decade of intensive research even to decide if the project was possible or not (Bloembergen et al. 1987, S9).

In 1988, the OTA followed up its 1985 with a second report on the feasibility of SDI. The agency pointed out that its initial report did not reflect any of the early research conducted by the newly created Strategic Defense Initiative Organization (SDIO), so contained a great deal of uncertainty. The 1988 report, however, was able to make use of three additional years of research by the SDIO to provide a better estimate of the prospects for the program. In this report, the OTA concluded that there was still a good deal of uncertainty as to whether the program could ever accomplish its goals, especially since the Congress had not appropriated nearly enough funds (nearly $15 billion) to carry out the background research needed. The agency suggested that under the very best of circumstances, with significantly increased funding, the SDI project might be feasible by the early decades of the 21st century (U.S. Congress, Office of Technology Assessment 1988, 3–5).

In the half decade following Reagan's 1983 speech, a number of other interested individuals and groups presented analyses of the likelihood that SDI would provide the protection that the president had hoped and expected from his dream project. The vast majority of those scientists and science-based groups expressed serious doubts as to whether the project could achieve those objectives, either ever or within the foreseeable future. A number of observers also expressed concerns that the program had been created and developed primarily on the basis of political, not scientific, issues. (See, for example, Bethe et al. 1984; Garwin 2013; Garwin et al. 1984; Glaser 1984; an excellent bibliography of articles in favor of and opposed to SDI can be found in Appendix J of the 1985 OTA report online at http://www.princeton.edu/~ota/disk2/1985/8504/850422.PDF.)

For all practical purposes, the SDI program began its demise in 1986, when Reagan refused to agree to limitations on the program in negotiations with Soviet president Mikhail Gorbochev over elimination of nuclear weapons in the two nations. Although research continued on elements of a missile defense system throughout Reagan's term in the presidency, SDI was finally abandoned in 1989, when President George H. W. Bush decided to focus on an element of the project that he called "brilliant pebbles." An SDI-like program in one form or another has continued to be a part of the U.S. defense system ever since. (For a review of the history of ballistic missile defense systems, see "US Ballistic Missile Defense Timeline: 1945–2008" 2013.) The most recent iteration of the program has also proved to be unreliable. A test of that system conducted in July 2013 failed when a missile launched from Vandenberg Air Force Base in California failed to intercept a target missile launched from the Marshall Islands in the Pacific Ocean. The failure was the third in succession in a test of a system to protect the U.S. West Coast against possible attack by North Korea ("Key U.S. Missile Interceptor Test Fails, Pentagon Says" 2013).

Science Interruptus

The three case studies presented here provide a glimpse of some of the ways in which politics and Big Science have interacted in the post–World War II era. In the case of the Manhattan Project, government essentially commandeered a major part of the scientific community and scientific activity in the United States in order to build nuclear weapons that allowed the nation to win World War II in a much shorter period of time than might otherwise have been possible. The authorization of the construction of the SSC illustrates the way in which proponents of basic research were able to convince the U.S. Congress, other decision makers in government, and the general public of the value of committing billions of dollars to the construction of a giant research machine, even though that project ultimately failed. And the debate over the SDI shows the conflicts that can arise when politicians and the government can push forward on an enormous military project, even when a large majority of professionals scientists counsel against such actions.

But the growth of science after World War II saw another side of the interaction of science and politics. Individuals and organizations from all parts of society began to reject science as an intellectual activity, the men and women who worked as scientists, the funding of science projects by federal and other governments, and the very results produced by scientific research. The last half century or more has seen an increasing amount of ignoring, denying, misinterpreting, hiding, and otherwise rejecting the knowledge produced by science. It is this rejection of science and scientists that the author has referred to by the phrase *science interruptus*. That process differs substantially from the traditional give and take between scientists and politicians over the way scientific research is to be used in a society and goes to the very legitimacy of the scientific enterprise itself.

The term *science interruptus* is used here to suggest a disruption in what might be considered the normal flow of information from scientific research to social utilization of that information. That is, scientific research often results in the production of new knowledge about some event in nature. That knowledge is then transmitted to social, political, economic, and other systems that use it (usually and hopefully) for the betterment of human society. In instances of *science interruptus*, that transmission is interrupted in some way or another, for some reason or another. Some examples of the way in which *science interruptus* may occur are the following:

- Ignorance of the research: Individuals both within and outside the scientific and political communities may simply be unaware that research on a particular topic has been conducted and new knowledge has been produced. A politician might make the argument, for example, that there are no technological fixes for global climate change because no research has as yet been conducted on that topic. Such is not the case because, in fact, a great deal of research has been done on the ways in which humans could make changes in Earth's ecosystem to counteract the effects of global warming. (See, for example, Inman 2013.) Anyone who does not mention this research (1) is not aware that the research has been conducted, (2) prefers not to acknowledge that the research has occurred and/or to recognize the potential applications of that research, usually because doing so will harm her or his argument about a topic, or (3) does not want her or his audience to be aware of the research and its potential practical consequences. (It is generally the case that a listener cannot tell which of these alternatives explains the speaker's failure to mention existing research.)

- Nonscientists sometimes acknowledge the existence of scientific research, but ridicule its validity. They suggest that scientists are wasting their time—and, more importantly, usually wasting taxpayer money—on research that has no conceivable value. One of the most famous examples of this position was a series of announcements released by Senator William Proxmire (D-Wisc.), beginning in 1975 and continuing for more than 13 years. Proxmire included in his monthly newsletter to constituents a particular piece of scientific research that he thought was a waste of taxpayer money. The 168 Golden Fleece Awards announced by Proxmire were given for seemingly absurd projects such as studying the effects on fish who drank tequila versus gin and a study as to whether people should carry umbrellas when it rains or not. Some, perhaps more, controversial awards were given for research on extraterrestrial life, facial recognition systems, and erotic effects from smoking marijuana ("Golden Fleece Awards, 1975–1987" 2013). The problems with Proxmire's approach were twofold. First, nonscientists may not always be the best judge as to which research projects are likely to have scientific value or not. Second, making fun of apparently silly scientific research may tend to produce more general negative attitudes toward research in general, thus precluding the development of some potentially useful new knowledge.

- Politicians may be very familiar with the scientific research on some topic of interest, but prefer to cite only specific parts of that research, a process that is sometimes known as *cherry-picking* of information. For example, there have probably been hundreds of research studies conducted in the past decade on the medical effects of smoking marijuana. The results of those studies cover a wide range, from harmful to beneficial, and everywhere in between. A person who wants to make a case for or against the smoking of marijuana for medical purposes can almost certainly find at least one study to support his or her case. But citing that one study (or a handful of studies) does not really provide convincing evidence for or against an argument.

- Another form of *science interruptus* occurs when a scientist or nonscientist chooses to invent new scientific information for which there is no experimental or observational basis. Practicing scientists resort to this practice on a surprisingly frequent basis, one that is, because of the nature of science, always discovered by colleagues. Nonscientists do so also, usually to strengthen a case for which they are arguing. Astronomer and science popularizer Carl Sagan once wrote about the United States' so-called war on drugs that "the government and munificently funded civic groups systematically distort and even invent scientific evidence of adverse effects (especially of marijuana)" (Sagan 1996, 390; this point is discussed in more detail in the section on marijuana that follows).

- A somewhat different form of *science interruptus* occurs when public officials simply stop the flow of scientific information they don't like by censoring or banning the publication of such information or by firing or not hiring researchers who have or who are likely to produce results with which officials are not comfortable. As just one of many possible examples of this practice is the report from October 2011 that the Texas Council on Environmental Quality (CEQ) decided to edit a report by Rice University oceanographer John Anderson on the state of Galveston Bay. CEQ officials decided to delete all references to possible effects of global climate change on conditions in the bay, changes that were of significance in understanding the current and future state of the bay (Piltz 2013).

In summary, the process of *science interruptus* may have all types of effects on the making of public policy, ranging from the relatively harmless (such as the Golden Fleece Awards) to the seriously harmful (such as the invention or misrepresentation of scientific information). The fact is, on some occasions, politicians simply lie to the general public, a practice that almost never results in sound public policy.

An important point to note about *science interruptus* in the modern world is that the practice is not a rarity, examples of which one has to search for; it occurs commonly in almost every major social issue for which there is a scientific component (which includes virtually all important social issues today). An example can be found in the career of one of the leading political figures in American politics in the first decade of the 21st century, Representative Michele Bachmann (R-Minn.).

Bachmann was a candidate for the Republican party nominee for president of the United States in 2011 and, at one point, was the most popular candidate among Republican voters in the campaign. The high point of her candidacy came in August 2011, when she won the Ames (Iowa) Straw Poll with 29 percent of the vote. She withdrew from the campaign five months later, however, when she came in sixth in the Iowa caucus elections (Henry 2013).

Bachmann became somewhat famous for a number of comments she made, many of which were about science-based topics and which often lacked factual accuracy. The following comments are examples. They have been rated for their factual accuracy by PoliticFact, an activity of the Tampa Bay Times, designed to check the accuracy of statements made by American politicians. Those statements are ranked on a scale from True, at one extreme, to False, almost at the other extreme, followed only by Pants on Fire, a term arising out of an old childhood saying, "Liar, liar, pants on fire." In each of the following, Bachmann's statement is followed by PolitiFact's assessment of its accuracy:

- "Scientists tell us that we could have a cure in 10 years for Alzheimer's" were it not for "overzealous regulators, excessive taxation and greedy litigators." Pants on Fire

- Of every "three dollars in food stamps for the needy, seven dollars in salaries and pensions (go to) the bureaucrats who are supposed to be taking care of the poor." Pants on Fire

- "One. That's the number of new drilling permits under the Obama administration since they came into office." Pants on Fire

- Page 92 of the House health care bill "says specifically that people can't purchase private health insurance after a date certain." Pants on Fire

- In the 1970s, "the swine flu broke out . . . under another Democrat, President Jimmy Carter." Pants on Fire

- "There's a woman who came up crying to me tonight after the debate. She said her daughter was given that [HPV] vaccine. She told me her daughter suffered mental retardation as a result. There are very dangerous consequences." False
- "The president released all of the oil from the Strategic Oil Reserve." False
- The government [under the Affordable Care Act] will "go out and buy my breast pump for my babies." False
- "The New England Journal of Medicine released a survey the week that President Obama signed Obamacare stating that over 30 percent of American physicians would leave the profession if the government took over health care." False
- Ezekiel Emanuel, one of President Obama's key health care advisers, "says, medical care should be reserved for the non-disabled. So watch out if you're disabled." False

Representative Bachmann announced in May 2013 that she would not run for reelection to the House of Representatives in 2014.

Opposition to scientific and technological advance is certainly not new in human history. During the early 18th century, for example, a movement developed in England based on opposition to technological developments that were depriving many individuals of their traditional means of livelihood. These individuals organized to carry out violent protests against the new technology that involved breaking and destroying the new machines, such as the spinning jenny, water frame, cotton gin, and spinning mule. Followers of this movement, known as *Luddites*, were concerned that the new technology would prevent them from earning a living (Jones 2006).

Nor has the Luddite movement completely died out. For example, the late 1990s saw the rise of a neo-Luddite movement organized to protest "consumerism and the increasingly bizarre and frightening technologies of the Computer Age." Like their predecessors, neo-Luddites believe in expressing their objections to modern technology with aggressive actions, including the sabotaging of "dam-building projects, deforestation, road building, electric power lines, tests on animals and other ecological crimes" (Sale 2013).

American history over the past decade is replete of examples of *science interruptus*. One glaring example is the early effort to ban the use of marijuana in the United States in the early 20th century.

The Evils of Marijuana

The marijuana plant (*Captiva spp.*) has been known to humans for thousands of years. Its earliest documented use dates to at least 12,000 BCE, at which time it was used as a fiber by a Neolithic tribe in China (Booth 2005, 20). Its three most common species, *Cannabis sativa*, *C. indica*, and *C. ruderalis*, have been used for a variety of purposes, primarily as medicinal products for treating a host of conditions, as a psychoactive drug, and, in the form of the fiber known as *hemp*, for the production of clothing, paper, ropes, and other types of cordage, sail canvas, netting, and oils (hemp oil).

Throughout history, the marijuana plant has been widely praised for its many beneficial uses and criticized for some of its less desirable effects on humans. The list of medical conditions for which cannabis has been used throughout history is so long that it might be easier just to list those conditions for which it has *not* been recommended. At one time or another, it has been used for the treatment of alcoholism, amyotophic lateral sclerosis (Lou Gehrig's disease), anorexia nervosa, arthritis, asthma, atherosclerosis, bipolar disorder, colorectal cancer, depression, dysenteria, dystonia, earache and ear diseases, epilepsy, digestive disorders, fever reduction, gout, hepatitis, inflammation, insomnia, leukemia, loss of appetite, low intelligence, malaria, memory loss, migraine, nausea and vomiting, neuralgia, premenstrual syndrome, skin tumors, pain of childbirth, Parkinson's disease, psoriasis, sickle-cell disease, sleep apnea, syphilis, weight loss, and the list goes on and on. Probably needless to say, modern medical research has by no means confirmed the value of cannabis for all or even most of these treatments (and, as far as some authorities are concerned, for *any* of the conditions listed).

The validity of marijuana as a medical treatment is illustrated by the fact that it has been listed in pharmacopeias throughout most of human history. The earliest such instance may date to about 1,500 BCE when the drug was listed in an early Chinese pharmacopeia, the *Rh Ya* (Brecher and the Editors of Consumer Reports 1972). The use of marijuana for medical purposes first appeared in the United States Pharmacopeia in 1850, where it was recommended for the treatment of alcoholism, anthrax, cholera, convulsive disorders, dysentery, excessive menstrual bleeding, gout, incontinence, insanity, leprosy, neuralgia, opiate addiction, rabies, tetanus, tonsillitis, and uterine bleeding, among other conditions (Boire and Feeney 2007, 16).

Very little is said in the historical record about the potentially harmful effects of using marijuana for medical purposes. However, concerns about its psychoactive effects do arise from time to time. Some medical workers, politicians, and others have expressed doubts because of the drug's tendency to produce both short- and long-term effects on the human brain and its normal functions. Books on the history of cannabis, medical marijuana, and related topics, however, generally fail to mention any significant, widespread prohibitions on the use of the drug until the early 20th century. (See, for example, Lee 2012.)

That situation began to change, however, in the early 20th century, at least partly because of a growing sense of moralism in American politics. In 1911, for example, Massachusetts became the first state to outlaw the sale and use of cannabis for any nonmedical purpose whatsoever. As was to be the case for at least a quarter century, the law was not passed in response to any expressed concerns about the dangers of the drug, but for the purpose of "completeness" in listing all known psychoactive products; the law was aimed primarily at the more dangerous substances, opium, morphine, and other narcotics (Gieringer 2013a). The Massachusetts law was also typical of other cannabis laws that were soon to come throughout the nation in that it made no effort to ban the drug, but to control its use by regulatory means (licenses, taxes, etc.) The Massachusetts law was soon followed by similar legislation in California, Maine, Wyoming, and Indiana (1913); New York City (1914); Utah and Vermont (1915); and Colorado and Nevada (1917). In all instances the laws were part of a puritanistic and regulatory fervor that attempted to impose morality through governmental control with legislation on not only cannabis, but also alcohol, oral sex, prizefighting, prostitution, and racetrack gambling (Gieringer 1999, 2013b).

The turning point in the campaign against the use of marijuana came in the early 1920s, especially at the Second Opium Conference and the International Opium Convention held in Geneva in February 1925. At that conference, the delegation of Egypt raised the issue of the harmful effects its nation was experiencing by individuals smoking hashish, a particularly potent form of cannabis. The delegates presented a document describing disastrous consequences of hashish use in the country. One delegate, Dr. Mohamed Abdel Salam El Guindy, for example, claimed that the use of hashish was responsible for most of the cases of insanity occurring in his country. He described the typical cannabis user as follows:

> His eye is wild and the expression of his face is stupid. He is silent; has no muscular power; suffers from physical ailments, heart troubles, digestive troubles etc; his intellectual faculties gradually weaken and the whole organism decays. The addict very frequently becomes neurasthenic and eventually insane. (United Nations Office on Drugs and Crime, Policy Analysis and Research Branch [2010], 54)

The author pauses at this point in his narrative for a brief, but important, commentary. Dr. El Guindy's comments come at a point in history when social views on the use of cannabis were generally favorable. Indeed, one of the first and most exhaustive early treatises on the medical uses of cannabis had been produced only 30 years earlier. In that report, Irish physician William Brooke O'Shaughnessy had conducted an extensive survey on the use of cannabis in India, where he was stationed with the British East India Company. He found that cannabis had a number of useful applications and very few serious drawbacks. O'Shaughnessy's work led to the creation by the British government of the Indian Hemp Drugs Commission in 1893 to obtain all available scientific information about the use of cannabis for medical purposes. The commission's nine-volume report, released in 1894, reached essentially the same conclusions as had O'Shaughnessy in his more limited research. By the time the Second Opium Conference and International Opium Conference were held in 1925, a voluminous scientific record attested to the safety and efficacy of cannabis in the treatment of a host of medical conditions. While Dr. El Guindy may very well have been correct in his concerns about hashish smoking in Egypt (he provided no data to support his general observations), conditions hardly appeared appropriate for a dramatic international revision of policies on the use of cannabis. And yet, that action is just what members at the conference took. They devoted two full chapters in their final report on the hazards of cannabis use (without having reviewed existing scientific evidence) and recommended the introduction of strict new standards on the import, export, sale, and use of cannabis (except for medical and research purposes). In one fell swoop, cannabis had become classified with some of the most dangerous drugs known, including heroin, morphine, opium, cocaine, and coca (Plouffe 2011). The tendency to criminalize cannabis, for all or some of its applications, without reference to scientific data, was to become standard procedure over much of the next century.

The turning point for the regulation of cannabis in the United States came in 1930 with the creation of the Federal Bureau of Narcotics (FBN). The FBN consolidated the functions of two earlier agencies, the Federal Narcotics Control Board and the Narcotics Division of the U.S. Department of the Treasury, both formed to deal primarily with opiate drugs, not marijuana. Under its new director, Harry J. Anslinger, however, the FBN began to work aggressively to monitor and control the growing, transport, sale, and use of cannabis products. At first, Anslinger and the FBN focused on its original target, the opiates. Before long, however, Anslinger began to turn his attention to what he perceived as the dangers of smoking marijuana. He began to write and speak about the terrible evils associated with the drug, producing a number of troubling comments about cannabis, such as:

- "A study of the effects of marihuana shows clearly that it is a dangerous drug, and Bureau records prove that its use is associated with insanity and crime. Therefore, from the standpoint of police work, it is a more dangerous drug than heroin or cocaine."

- "Narcotic effects are good or bad. Marihuana effects run in one direction only, and that is bad. Marihuana weakens the will."

- "I believe in some cases one cigarette might develop a homicidal mania, probably to kill his brother. . . . Probably some people could smoke five before it would take effect, but all the experts agree that the continued use leads to insanity."

- "How many murders, suicides, robberies, criminal assaults, holdups, burglaries, and deeds of maniacal insanity it causes each year, especially among the young, can be only conjectured. . . . No one knows, when he places a marijuana cigarette to his lips, whether he will become a philosopher, a joyous reveler in a musical heaven, a mad insensate, a calm philosopher, or a murderer." ("Harry J. Anslinger Quotes/

Quotations" 2013. These quotations have confirmed citations. A number of other comments attributed to Anslinger
are available, although confirmed citations are not always
provided. See, for example, "Harry J. Anslinger Quotes"
2013. For Anslinger's comments before the U.S. Congress,
also see "Taxation of Marihuana" 2013.)

Although Anslinger is usually nominated as the chief "vil-
lain" in criminalizing the use of marijuana, some evidence
suggests that other important figures were also active in the
campaign. For example, newspaper publisher William Ran-
dolph Hearst also carried on a vitriolic campaign against cannabis in his papers, supposedly because he was concerned that
the hemp form of the plant would displace pulp from his vast
timber holdings. The Dupont family's chemical company was
also accused of attempting to ban the growth of hemp because
of its potential competition for the company's newly discovered nylon products (McCabe 2011, 149–163).

In any case, the attempt to vilify the use of marijuana in the
United States was eventually successful, resulting in the adoption of the Marihuana Tax Act of 1937 (with the spelling of the
word popular at the time). The act did not actually criminalize
the use of marijuana, although it imposed tax restrictions that
made the drug's use illegal for all practical purposes. Within
a week of the law's adoption, federal agents had arrested two
men in Denver, Colorado, for possession (in one case) and the
sale (in the other case) of marijuana without the necessary tax
stamps. The arrests resulted in prison terms of 18 months and 4
years, respectively, the first of the government's long history of
prosecutions for possession and sale of the drug ("U.S. District
Court, Denver, Colorado Imposes First Federal Marihuana
Law Penalties" 2013).

The interesting point about this nascent period in the history of marijuana laws is the extent to which Anslinger, Hearst,
and their colleagues "massaged" the facts about marijuana use.
Whether one agrees or disagrees about the legal status of the

drug, the fact is that many of the statements made by the proponents of criminalization were factually inaccurate. It is somewhat striking that, among the many individuals who appeared before the Congressional committee considering the bill, only one, Dr. William C. Woodward, a representative of the American Medical Association (AMA), spoke in opposition to the bill. He pointed out that there was no scientific evidence suggesting that marijuana is a harmful drug, so there was no reason to ban its manufacture and use.

Woodward's testimony was apparently not what the committee wanted to hear. During one of Woodward's appearances before the committee, one Congressman observed that "Doctor, if you can't say something good about what we are trying to do, why don't you go home?" Immediately following that observation, a second member said "Doctor, if you haven't got something better to say than that, we are sick of hearing you" (Whitebread 2013). As these comments might suggest, the committee went ahead and recommended adoption of the Marihuana Tax Act.

Passage of that act by no means ended the controversy over the safety and efficacy for medical purposes of the use of cannabis products. One of the most immediate responses to passage of the act was the commissioning of a study on the effects of smoking marijuana by New York City mayor Fiorello La Guardia. That study, conducted by the New York Academy of Medicine, reported a number of results diametrically opposed to those presented to the Congressional committee by Anslinger and other proponents of criminalization of the drug. For example, the report said that the use of marijuana does not lead to the use of more dangerous drugs (the so-called *gateway theory*), is not addictive, is not involved in the commission of major crimes, and is not particularly common among young people in the city ("The La Guardia Report—Sociological Study" 2013).

Anslinger was furious about the committee's report and decided to take every step necessary to discredit the report and even to prevent its publication. He was not able to do the latter, although he did threaten to arrest and imprison any medical or scientific researcher who attempted to conduct experiments on cannabis products (Herer 2001). Anslinger's policy about marijuana research has, however, been one of his most lasting legacies in the field. Because of its control over the legal supply of the drug, the U.S. Drug Enforcement Administration (DEA) continues to determine when, how, and what research on the drug will be allowed and, for the most part, the qualifiers for those terms are "never," "not," and "none" (Ferro 2013).

The scientific and political controversy over the use of marijuana that began in the 1930s continues in the United States today. The U.S. government's attitude toward the drug became solidified in the adoption of the Controlled Substances Act of 1970 (CSA), in which the U.S. Congress created five levels (schedules) of controlled substances. Placement of a drug is determined by three criteria: availability of any generally accepted medical use, safety of use under medical supervision, and potential for abuse. For example, drugs placed in Schedule I are all listed in that category because they have no legitimate medical use, are not safe to use even under medical supervision, and have a high risk of abuse by users ("Controlled Substances Schedules" 2013). Among the drugs listed in Schedule I are heroin, lysergic acid diethylamide (LSD), methaqualone, morphine, 3,4-methylenedioxymethamphetamine (Ecstasy), peyote, and marijuana (cannabis).

One more effort was made to find and express scientific consensus on the efficacy and safety of marijuana. When the U.S. Congress passed the CSA, it provided funding for a commission to investigate the status of marijuana, with a view toward deciding the schedule to which it should be assigned. President Richard M. Nixon appointed Governor Raymond P. Shafer (R-Pa.) to head the committee, reminding the governor that "you're enough of a 'pro' to know that for you to come out with something that would run counter to what the Congress feels and what the country feels and what we're planning to do, would make your Commission just look bad as hell" ("Nixon Tapes Show Roots of Marijuana Prohibition: Misinformation, Culture Wars and Prejudice" 2013).

The commission's report, *Marihuana: A Signal of Misunderstanding* (also called the Shafer report), proved to be a huge disappointment to Nixon and other supporters of the prohibition of marijuana. It found that concerns about the physical, mental, emotional, psychological, and social effects of the drug were grossly exaggerated, and that legislation should focus on preventing heavy, long-term use rather than punishment of short-term and/or intermittent use of small amounts of the product ("The Report of the National Commission on Marihuana and Drug Abuse" 2013). In the end, Nixon simply ignored the findings of the Shafer Commission and made an aggressive "war on drugs" a key feature of his 1972 campaign for reelection (which he won in a landslide).

Neither did the Shafer report resolve the issue as to where marijuana should be placed on the schedule of drugs. It simply ended up as a Schedule I drug. The question over which many individuals and organizations have battled for nearly half a century is whether marijuana belongs in this drug category, or whether it should be listed in a less restrictive class. After all, some commentators observe, the drug has been used for medical purposes for many centuries without the type of harm to humans associated with other Schedule I drugs. As of early 2014, however, the U.S. government has made no effort no reclassify the drug, no matter what the scientific and medical evidence might say about the issue. The reason appears to be that the political consequences accompanying such a change would be too severe for the government to bear (Scannell 2013).

And yet, as this debate continues, a solid majority of Americans appear to support the use of marijuana for medical purposes. The Gallup poll organization has found that about three-quarters of all Americans polled support the use of marijuana for medical purposes, a fraction that has not significantly changed between 1999, when the question was first asked, to

2010, the most recent poll available ("Illegal Drugs" 2013). Over that period of time, 18 states and the District of Columbia have adopted policies that permit the use of medical marijuana within their boundaries ("Medical Marijuana" 2013).

Perhaps even more significantly, individual states have begun to consider and, in some cases, approve the use of small amounts of marijuana for nonmedical purposes. The first two states to take such action were Colorado and Washington, both of which adopted laws in 2012 permitting possession of small amounts of the drug. The trend in this direction was highlighted in June 2013, however, when the U.S. Conference of Mayors unanimously adopted a resolution asking the federal government to respect state and local laws permitting the possession of small amounts of marijuana for medical or nonmedical purposes (Altieri 2013).

In the United States in the 2010s, federal, state, and local governments are taking very different views of the scientific evidence about the use of marijuana laws and using the evidence to come to very different policy decisions for their own constituencies. The next big question will be what happens when those views come into conflict.

Issues of Human Sexuality

One reason that marijuana has long been such a contentious issue in the United States (and other parts of the world) probably is that so many people have such strong feelings, both pro and con, about the drug. When individuals are passionate about a topic, they may feel compelled to hide, misinterpret, invent, or deny factual information about the topic in order to present a stronger case for their side of the argument. The evidence from tapes made during the presidency of Richard M. Nixon make it very clear that he felt so strongly about the dangers of marijuana as a drug that he was willing to use (or ignore) facts in whatever ways were necessary to produce the public policy he felt was essential for the American people. (See

"Nixon Tapes Show Roots of Marijuana Prohibition: Misinformation, Culture Wars and Prejudice" 2013 for more details on this point.)

Another field in which emotions run high—and the potential for the shading of factual information is also great—is that of human sexuality. Perhaps no aspect of human life is so fraught with religious, ethical, moral, and social content the way in which men and women interact with each other sexually. It is to be expected, then, that government policy on sexual issues is almost inevitably loaded with nonscientific factors in making scientific decisions. Of the many topics that might be included in this area, one that received particular attention over the past three decades has been abstinence-only education. The general objective of abstinence-only education is that young men and women should avoid all sexual contact until they are married. Among the many purported advantages of abstinence-only education (AOE) is a sharp reduction in the number of unplanned pregnancies and sexually transmitted infections (STIs).

Abstinence-Only Education

Largely a phenomenon of the American educational system, AOE first received federal recognition in the 1981 Adolescent Family Life Act (AFLA), Title XX of the Public Health Service Act. The act provided funding in the amount of $11.08 million to the Office of Adolescent Pregnancy Prevention of the Department of Health and Human Services to create family-centered programs for the purposes of promoting "chastity and self-discipline" (Saul 2013; "Title XX Funding History" 2013). At the time, the legislation was based on no scientific evidence that AOE was effective in producing such a result; it instead reflected the religious and social beliefs of its legislative sponsors, senators Jeremiah Denton (R-Ala.) and Orrin Hatch (R-Utah).

AFLA was expanded in 1996 with the passage of Title V of the Welfare Reform Act, also known as the Temporary

Assistance for Needy Families Act (TANFA). That act provided for a new system of AOE grants to the states that were based on an eight-part definition for abstinence-only programs. The definition became known as the *A-H definition*, because of its labeling in the act. The eight definitions were not based on scientific criteria, but on social, religious, and psychological criteria, such as:

- the exclusive purpose of teaching the social, psychological, and health gains to be realized by abstaining from sexual activity;
- a mutually faithful monogamous relationship in the context of marriage is the expected standard of sexual activity;
- bearing children out-of-wedlock is likely to have harmful consequences for the child, the child's parents, and society; and
- abstinence from sexual activity is the only certain way to avoid out-of-wedlock pregnancy, sexually transmitted diseases, and other associated health problems. ("H.R. 3734" 2013, Sec. 912)

Adoption of TANFA doubled the amount of funding available nationally for AOE programs, from $7.7 million in 1996 to $14.9 million in 1997. Funding increased again in 2001, following the establishment of a third system of AEO grants with an amendment to Title XI, section 1110 of the Social Security Act by Congress in 2000. By 2003, federal funding for all forms of AEO had reached a total of $30.9 million, a level at which it remained until 2009, when President Barack Obama began to cut funding to the programs by about half ("Title XX Funding History" 2013).

By the turn of the century, a number of individuals and organizations were asking how effective AOE programs were in achieving the goals for which they were created. Part of this concern was based on the fact that the U.S. government had already spent nearly half a billion dollars on such programs, with little or no evaluation as to their effectiveness. The first such report on AOE programs was issued by a special committee of the House Committee on Government Reform, headed by representative Henry Waxman (D-Calif.) in December 2004. The report studied 13 programs receiving the largest federal funding to determine their effectiveness in accomplishing goals they had established for themselves. That study found that more than 80 percent of the curricula developed by these programs contained "false, misleading, or distorted information about reproductive health" (United States House of Representatives. Committee on Government Reform—Minority Staff. Special Investigations Division 2013, I).

The nonpartisan Government Accountability Office (GAO) also conducted a study on the effectiveness of AOE programs in 2006 (Abstinence Education 2006). The GAO was unable to come to any very specific conclusions because, as it noted, the granting agency for AOE programs, the Office of Population Affairs (OPA), did not require grantees to design or conduct any type of evaluation for their programs. The GAO also found that evaluation studies conducted by states did not meet even minimal standards for obtaining objective data on such programs (Abstinence Education 2006, 6).

A third study was authorized by the Congress in 1997 as a means of evaluating the success of federally funded AOE programs. That study was conducted by Mathematica Policy Research, Inc., of Princeton, New Jersey. Mathematica selected four AOE programs geographically distributed throughout the United States. The study resulted in two major conclusions. First, students who had taken AOE classes in one of the four programs were no more likely to have remained abstinent than were their peers in control groups. Second, students in both study and control groups were equally likely to have engaged in unprotected sex in both short and long terms (12 months) during the study. The study also measured knowledge and attitudes about a number of sex-related issues, such as use of condoms and STIs. Researchers found modest differences in one direction or another on most of these measures, differences that did not provide strong support for one form of sex education (such as AOE) over another (Trenholm et al. 2007).

A number of states have also conducted evaluation studies of AOE programs within their jurisdiction. These studies tend to show little or no positive effect of the programs and, in some cases, some harmful results. For example:

- Pennsylvania found the programs "largely ineffective in reducing sexual onset and promoting attitudes and skills consistent with sexual abstinence";

- Texas found "no significant changes" in sexual behaviors as a result of taking an AOE program;

- The Arizona report said that "sexual behavior rates do not appear to be changing";

- A Kansas study found "no changes noted for participants' actual or intended behavior; such as whether they planned to wait until marriage to have sex";

- An independent study in Minnesota found that "sexual activity doubled among junior high school participants" in a state AOE program. ("What the Research Says . . . Abstinence-Only-Until-Marriage Programs" 2013)

These results appear to have had a depressing effect on the willingness of the federal government to continue funding AOE programs. In his 2012 budget, for example, President Barack Obama set aside only $5 million for such programs. The Republican-controlled House of Representatives increased that amount 10-fold, however, to $50 million (Koebler 2013).

The point of the discussion about AOE is that federal, state, and local governments have appeared to be willing to spend vast amounts of money—as much as $1.5 billion between 1996 and 2009, according to some estimates—for the teaching of a program in human sexuality that apparently is essentially ineffective in achieving the goals for which it was designed (Koebler 2013).

As recently as 2011, more than half (21 states) of the 38 states that required some form of sex education in local schools emphasized AOE programs over all other forms of sex education either by law or policy. (The 21 states are Alabama, Arizona, Arkansas, Delaware, Florida, Hawaii, Illinois, Indiana, Kentucky, Louisiana, Mississippi, New Mexico, North Carolina, Ohio, Oklahoma, Pennsylvania, Rhode Island, South Carolina, Tennessee, Texas, and Utah.) A study of these state laws and pregnancy rates found that within the states that stressed AOE most strongly, "abstinence-only education does not reduce and likely increases teen pregnancy rates" (Stranger-Hall and Hall 2013).

Plan B

Of the many examples of the conflict between science and politics one might find in the realm of reproductive health, one of the most prolonged, contentious, and most recent has been the battle over the use of a form of emergency contraception called *Plan B*. The term *emergency contraceptive* refers to a form of birth control used after a woman has engaged in unprotected sexual activity, such as may occur as the result of rape or consensual sexual activity without the use of other forms of birth control. Emergency contraceptives are also known as *morning-after pills* and *post-coital contraceptives*.

Plan B is the brand name for a substance produced by Duramed Pharmaceuticals that contains the hormone levonorgestrel. Levonorgestrel was first synthesized by American chemist Herschel Smith of Wyeth Pharmaceuticals in the early 1960s. It was first used as a contraceptive in the 1980s as an implantable contraceptive under the commercial name of Norplant. It was eventually replaced by other hormones for that purpose, but continued to find use as a coating on intrauterine contraceptive devices (IUDs). By the late 1990s, Duramed had developed a contraceptive pill, which they called Plan B, which prevents pregnancy for up to 120 hours after intercourse, with rapidly decreasing effectiveness toward the end of that period.

In 1999, the U.S. Food and Drug Administration (FDA) issued approval for the use of Plan B as a prescription contraceptive for women. Almost immediately, Teva Branded Pharmaceuticals (an offshoot of Duramed; the company's name continued to change during the history of this issue) and a number of women's health groups filed a petition with the FDA to allow the sale of Plan B over the counter, without prescription. As is normally the case, the FDA submitted the petition for review, this time to a joint session of its Reproductive Health Drugs Advisory Committee and Non-Prescription Drugs Advisory Committee. The joint committee held its first meeting on December 16, 2003, and the history of the approval process became a complex mix of science and politics from that point on.

In the first place, members of the administration of President George W. Bush had learned of the petition to open Plan B use to over-the-counter (OTC) status, and decided to add three new members to one of the committees about to debate the merits of the proposal. As one court document on the topic later reported, "despite the usual practice by which members of the advisory committees are proposed by career scientists with relevant expertise, these three members were proposed by political appointees. Each of the three voted contrary to the majority vote of the Committee on one or more votes taken regarding Plan B" (In the United States District Court For the Eastern District of New York 2013, §65). At least one of the three new appointees had a long record of taking what might be described as a somewhat nontraditional view with regard to women's reproductive health, recommending religious rather than purely scientific approaches to some problems in this area. (For more details, see McGarvey 2005.)

In any case, the joint committee eventually voted 23 to 4 to recommend that Plan B be made available OTC. At that point, supporters of the action were jubilant, because the FDA rarely acts in opposition to the recommendations of its advisory committees. They were in for a surprise in this case, however, as

FDA officials responsible for taking this action decided instead to ignore the committees' advice and to reject the petition. This action was later found to have been based not on any scientific information, but on concerns by the officials that they might lose their jobs if they approved the petition. For example, Dr. Steven Galson, then acting director of the FDA Center for Drug Evaluation and Research (CDER), told a colleague that he feared "his career at the FDA would be jeopardized" if he approved the Plan B petition (In the United States District Court For the Eastern District of New York 2013, §85). Second-in-command at the FDA, Dr. Janet Woodcock, also noted that rejection of the petition was essential as it was "the only course of action that would appease the Presidential administration's constituents" (In the United States District Court For the Eastern District of New York 2013, §74; also see "The Fight for Emergency Contraception: Every Second Counts" 2013).

The debate over Plan B continued for most of the next decade, with the FDA continuing to delay action on earlier petitions for its use as an OTC drug and citizen groups pushing the government (usually in court cases) to take such an action. These groups continued to point out that the drug had been extensively studied by medical experts and cleared by them for use by all females of whatever age, with virtually no risk to users. Finally, in August 2006, the FDA relented and agreed to allow the sale of Plan B as an OTC product, but only if it were made available "behind the counter," that is, not on pharmacy shelves but only by request of the pharmacist, and only for women aged 18 and older ("The Fight for Emergency Contraception: Every Second Counts" 2013).

Women's health groups were still not satisfied with this decision, and sued the FDA one more time to get approval of the drug for women of all ages. (Also, by this time the battle was being fought over various forms of the pill that had been developed since the controversy began a decade earlier.) However, even a change in presidential administrations, from that of Republican president George W. Bush to Democratic president

Barack Obama did not change the FDA's stand on Plan B. In his first year of the Obama presidency (2009), the FDA did lower the age at which the drug was available to women to 17, but went no further in responding the requests to make it available to all women of any age.

Finally, in 2011, largely as the result of court decisions directing it to do so, the FDA announced that it would finally approve the 2001 petition asking that Plan B be made available to all women of every age without prescription. FDA administrator Dr. Margaret Hamburg announced the new policy, saying that "there is adequate and reasonable, well-supported, and science-based evidence that Plan B One-Step is safe and effective and should be approved for nonprescription use for all females of child-bearing potential" ("Commissioner Statement" 2013).

Hamburg's decision was overturned, however, by Secretary of Health and Human Services (HHS) Kathleen Sebelius, who responded to the FDA decision by saying that she did not believe that "enough data were presented" to support the request for OTC sale of the drug to women of all ages ("Statement by U.S. Department of Health and Human Services Secretary Kathleen Sebelius" 2013). Her decision left in place the 2009 FDA decision, a move that was supported the next day by Obama himself, who noted that he was the father of two young women, and that it was important to "apply some common sense to various rules when it comes to over-the-counter medicine" ("Statement by the President" 2013).

The long battle over Plan B finally came to an end in 2013, when Judge Edward Korman of the Eastern District Court of New York ruled that the FDA was required to make the drug available to all women of any age without prescription. In his decision, Judge Korman excoriated the FDA and the Obama administration for making a series of decisions about Plan B based on political considerations rather than scientific evidence. He called their actions "arbitrary, capricious, and unreasonable," and noted that Secretary Sebelius' actions in 2011 were "politically motivated, scientifically unjustified, and contrary to agency precedent" (Belluck 2013). (In fact, although Sebelius was entirely within her legal rights to overturn Hamburg's decision, such an action had never before been taken by a secretary of HHS.)

Somewhat remarkably, the FDA responded to Judge Korman's ruling by announcing on April 30 that it was approving the use of Plan B for women age 15 and older! That is, the agency essentially ignored the judge's order and responded only by lowering the age of permitted use by two years. When women's health groups once more objected to this tactic, the FDA and Obama administration finally "caved in" on June 20, 2013, announcing that is was complying with Judge Korman's order of two months earlier ("FDA Approves Plan B One-step Emergency Contraceptive for Use without a Prescription for All Women of Child-bearing Potential" 2013).

Global Climate Change

On February 18, 2004, 62 leading American scientists signed a petition expressing concern about the misuse of scientific information by the administration of President George W. Bush. Over the following four years, the list of signers reached more than 15,000 names, including 48 Nobel laureates, 62 recipients of the National Medal of Science, and 127 members of the National Academy of Sciences (Scientific Integrity in Policy Making 2004, 6). A month after the petition was issued, the Union of Concerned Scientists (UCS) released a report detailing its concerns about the misuses of science by the Bush administration in a number of areas, including climate change, health research, HIV/AIDS issues, air quality, endangered species, and stem cell research. That report was later expanded, revised, and published in a popular form by lead researcher Seth Shulman in the book *Undermining Science: Suppression and Distortion in the Bush Administration* (Shulman 2004). The report's discussion of the way in which climate change issues were handled by the Bush administration is particularly instructive for this book.

Global climate change is one of the most complex of current science-based political issues. A number of excellent reference books are available for those readers who are interested in background information on the topic, including a number listed in Chapter 6 of this book. (See especially Emanuel 2012; Fletcher 2013; Peters 2013.)

The issue is based on evidence that Earth's climate has been warming over at least the last half century, and probably much longer. The most fundamental aspects of this change have now been verified by scientific research, and there is virtually no disagreement among scientists or nonscientists about the warming phenomenon itself. That is, just about everyone agrees that Earth's average annual temperature is increasing and that certain chemicals in the atmosphere known as *greenhouse gases* (such as methane and carbon dioxide) are also increasing at a regular rate. There is less unanimity about the potential long-term effects of these changes, although the vast majority of scientists believe that, if they continue, such changes are likely to make profound changes in Earth's climate that will raise a host of political, social, economic, agricultural health, and other issues. The nature and extent of these issues are difficult to predict, given the very large scope of the problem (the whole planet) and the challenge of predicting changes that may not occur for decades or longer. Nonetheless, the great majority of professional scientists who study climate are convinced that, absent significant changes in human activities responsible for the release of greenhouse gases into the atmosphere, climate changes are likely to be responsible for a number of very significant problems for humans beginning in the next half century or so. (Of the many predictions about possible effects from global climate change, probably the most comprehensive and responsible are those issued by the Intergovernmental Panel on Climate Change, IPCC. See http://www.ipcc.ch/. For a useful general discussion of this issue, also see "Global Climate Change" 2013.)

The political question raised by this information is what steps nations should take, if any, to reduce global warming and to ameliorate and/or adjust for its current and future effects, if anything. The problem is that reducing greenhouse gas emissions almost certainly involves making significant changes in the way human societies operate. For example, the largest single source of greenhouse gases is the release of carbon dioxide into the atmosphere by the combustion of fossil fuels. But reducing the release of carbon dioxide probably means cutting back on many essential industrial operations (such as energy production) that are fundamental elements of modern developed societies. It is difficult to imagine a politician in the 21st century, for example, telling the president of a major international oil company that the world will have to start living with less petroleum extraction and combustion over coming years, or explaining to constituents that they may have to pay twice as much for a gallon of gasoline to deal with climate issues.

Among the many ways in which politicians and government officials can deal with issues of global climate change is *science interruptus*. That is, politicians and government officials can hide, ignore, deny, lie and mislead about, quote selectively from, suppress, and otherwise interfere with the normal flow of scientific information about climate change to decision-making groups and individuals and the general public. (It goes without saying that there are endless ways in which politicians and government officials can also make honest efforts to use existing scientific information to develop helpful and useful public policy about climate change.) The point of the following section, then, is not to argue one side or the other of the ongoing debate over climate change, but to describe some of the less savory methods that politicians have used to prevent a legitimate discussion of the issue.

As a candidate for president in the 2000 election, George W. Bush expressed concern about climate change and called it a problem that "must be addressed by the world" because it had "the potential to impact every corner of the world" (http://

www.intellectualtakeout.org/library/primary-sources/george-w-bush-remarks-global-climate-change). Based on such comments, environmentalists perhaps had some hope that Bush would address global climate change issues aggressively if elected President. Such an event did not occur, however, as the Bush administration was marked almost from its beginning by efforts to suppress information about climate change and deny the severity of climate change issues in the United States.

As an example, the organization primarily responsible for studying climate change issues internationally, the Intergovernmental Panel on Climate Change (IPCC), issued an important report on the status of the issue, the Third Assessment Report on Climate Change, in January 2001, only a few days after Bush's inauguration as president. Apparently dissatisfied with the content of the IPCC report, Bush directed the U.S. National Academy of Sciences (NAS) to conduct a review of that report, focusing particularly on uncertainties in scientists' understanding of the climate change process and consequences. The NAS issued its report in July 2001, confirming, for the most part, the conclusions in the IPCC report (Oreskes 2007, 68).

A year later, yet another important report on climate change was released, this time by the U.S. Environmental Protection Agency (EPA). The report, the National Assessment of Climate Change Impacts (NCA), had been mandated by a United Nations agreement to which the United States was a signatory. It reached essentially the same conclusions about the causes and likely consequences of global climate change as those in the IPCC and NAS documents (U.S. Climate Action Report—2002). President Bush dismissed the information contained in the report, noting simply that it was only a "report put out by the bureaucracy." ("Update on Climate Change Policy" 2013; also see "Global Climate Change and the U.S. Climate Action Report" 2013.)

Such treatment of scientific information on climate change by the president was probably totally within his rights in deciding what information to use and which to ignore in making public policy decisions. But members of his administration went well beyond this approach to climate change issues. One prominent example of this situation was the work of Philip Cooney, then chief of staff at the Council for Environmental Quality (CEQ). President Bush appointed Cooney to this position in 2001, after he had served for many years as a lobbyist for the American Petroleum Institute. As chief of staff of the CEQ, Cooney undertook a campaign to reduce U.S. government attention to and concerns about global warming. According to Rick Piltz, then a member of the U.S. Global Change Research Program, Cooney began a program of "political policing" that included orders to refrain from making any reference to the NCA report and editing of CEQ reports to downplay the potential effects of climate change on the American economy ("Interviews: Rick Piltz" 2013). Again, Cooney was well within his rights to edit documents produced by the CEQ, although he personally had essentially no background in science himself that would have qualified him to make some of the judgments he apparently made ("RPTS Scott" 2013, 15; See also Revkin 2013a). Cooney resigned from his post at CEQ two days after news of his editing of documents had become widely known. A few days later, Cooney was hired by Exxon Mobil for a position with "unspecified responsibilities" (Revkin 2013b).

One might hope that the Cooney experience with climate change science was a unique or rare experience for scientists working in and for the government. But a 2007 report by the UCS and the Government Accountability Project (GAP) found that that episode was surprisingly common within the climate change research community. That report noted that,

- nearly half (46 percent) of all respondents had experienced pressure to delete the terms *global warming* or *climate change* from their reports;
- two in five (43 percent) had perceived or experienced editorial changes that altered the scientific meaning of their findings;

- nearly half (46 percent) had experienced changes in work conditions that interfered with their climate-related work;
- a quarter (25 percent) knew of or personally experienced situations in which scientists objected to or removed themselves from research projects as a result of pressures to change their scientific findings;
- 150 respondents (58 percent) had personally experienced one or more situations of political interference with their work in the preceding five years. (Donaghy et al. 2007, 2)

Additional studies by the UCS have found that climate change researchers are not unique in the political pressures they feel from political and governmental agencies and individuals. In a series of studies, the organization has reported on such pressures on researchers working in and for the Environmental Protection agency (EPA; Interference at the EPA 2008), the Food and Drug Administration (FDA; "Voices of Scientists at FDA: Protecting Public Health Depends on Independent Science" 2013), the Fish and Wildlife Service (FWS; "U.S. Fish & Wildlife Service Survey Summary. February 2005" 2013), and the National Oceanic and Atmospheric Administration (NOAA; "Survey of NOAA Fisheries Service Employees" 2013). During the first decade of the 21st century, political interference with scientific research had clearly become a common occurrence in the United States. And what does that have to say about the way in which scientific information is to be collected and used in the country?

Conclusion

Science and politics have had a long, complex, and interwoven history throughout the world as well as in the United States. In some instances, that interrelationship has had a strongly beneficial effect on one partner or the other and, in some cases, on both. In other cases, it has proved to be harmful to one partner or the other, and in some instances, to both. Perhaps the most important lesson that can be learned from this long history is that all parties are probably best served in the end when the pressures brought by scientists upon politicians and governmental officials, as well as by politics and government on science, are clearly evident to all parties involved, as well as to the general public, whenever such pressures are brought to bear.

References

Abstinence Education. 2006. Washington, DC: General Accountability Office. Available online at http://www.gao .gov/new.items/d0787.pdf. Accessed on June 27, 2013.

Alsop, Joseph. 1959. "True Missile Gap Picture Belies Pentagon Response." *Eugene Register-Guard*, October 13, 1959, 10A. http://news.google.com/newspapers?nid=1310&dat =19591013&id=5xZWAAAAIBAJ&sjid=8eIDAAAAIBAJ &pg=7129,2270259. Accessed on June 7, 2013.

Alsop, Joseph, and Stewart Alsop. 1954. "Reliability of Ex-Communist 'Expert Witness' Questioned." *Toledo Blade*. http://news.google.com/newspapers?nid=1350&dat= 19540416&id=c_ESAAAAIBAJ&sjid=ggAEAAAAIBAJ &pg=6039,138285. Accessed on June 3, 2013.

Altieri, Eric. "US Conference of Mayors Unanimously Pass Resolution Calling for the Feds to Respect Local Marijuana Laws." NORML, http://blog.norml.org/2013/06/24/ us-conference-of-mayors-unanimously-pass-resolution-calling-for-the-feds-to-respect-local-marijuana-laws/. Accessed on June 24, 2013.

"America Starts for the Moon: 1957–1963." NASA History Program Office. http://www.hq.nasa.gov/pao/History/ SP-4214/ch1-4.html. Accessed on June 7, 2013.

"The Apollo Program." National Aeronautics and Space Administration. http://spaceflight.nasa.gov/history/apollo/. Accessed on June 9, 2013.

Bazell, Robert. 1987. "Quark Barrel Politics." *New Republic*. 196(25) (June 22, 1987): 9-10.

Belluck, Pam. "Judge Strikes Down Age Limits on Morning-after Pill." *New York Times.* April 5, 2013. Available online at http://www.nytimes.com/2013/04/06/health/judge-orders-fda-to-make-morning-after-pill-available-over-the-counter-for-all-ages.html?ref=planbcontraceptive. Accessed on June 30, 2013.

Bethe, Hans A., et al. 1984. "Space-based Ballistic Missile Defense." *Scientific American.* 251(4) (October 1984): 39–49.

Bimber, Bruce. 1996. *Politics of Expertise in Congress: The Rise and Fall of the Office of Technology Assessment.* Albany: State University of New York Press.

Bird, Kai, and Martin J. Sherwin. 2005. *American Prometheus: The Triumph and Tragedy of J. Robert Oppenheimer.* New York: A. A. Knopf.

Bloembergen, N., et al. 1987. "Report to the APS of the Study Group on Science and Technology of Directed Energy Weapons." *Reviews of Modern Physics.* 59(3, II): S1–S168.

Boesman, William C. *Issue Brief.* Updated August 27, 1987. Washington, DC: Congressional Research Service, 1987. Available online at http://digital.library.unt.edu/ark:/67531/metacrs8880/m1/1/high_res_d/IB87096_1987Aug27.pdf. Accessed on June 9, 2013.

Boire, Richard Glen, and Kevin Feeney. 2007. *Medical Marijuana Law.* Berkeley: Ronin.

Booth, Martin. *Cannabis: A History.* New York: Picador Press, 2005.

Brecher, Edward M., and the Editors of Consumer Reports. 1972. *Licit and Illicit Drugs.* Mount Vernon, NY: Consumers Union. Available online at http://www.druglibrary.org/schaffer/library/studies/cu/cu53.html. Accessed on June 24, 2013.

Burress, Charles. "The Oppenheimer Riddle: New Evidence of Communist Membership Debated by Scholars of Berkeley Scientist." SFGate. http://www.sfgate.com/politics/

article/THE-OPPENHEIMER-RIDDLE-New-evidence-of-2788653.php. Accessed on June 4, 2013.

Carson, Cathryn, and David A. Hollinger. 2005. *Reappraising Oppenheimer: Centennial Studies and Reflections*. Berkeley: Office for History of Science and Technology, University of California.

"Commissioner Statement." U.S. Food and Drug Administration. http://www.fda.gov/NewsEvents/Newsroom/ucm282805.htm. Accessed on June 30, 2013.

"Controlled Substances Schedules." U.S. Department of Justice. Drug Enforcement Administration. Office of Diversion Control. http://www.deadiversion.usdoj.gov/schedules/. Accessed on June 26, 2013.

Day, Dwayne A. "Of Myths and Missiles: The Truth about John F. Kennedy and the Missile Gap." The Space Review. http://www.thespacereview.com/article/523/1. Accessed on June 7, 2013.

De Santillana, Giorgio. 1955. *The Crime of Galileo*. Chicago: University of Chicago Press.

Donaghy, Timothy, et al. 2007. *Atmosphere of Pressure: Political Interference in Federal Climate Science*. Union of Concerned Scientists and Government Accountability Project, February 2007. http://www.whistleblower.org/storage/documents/AtmosphereOfPressure.pdf. Accessed on July 21, 2013.

Emanuel, Kerry. 2012. *What We Know about Climate Change*, 2nd ed. Cambridge, MA: MIT Press.

"FDA Approves Plan B One-step Emergency Contraceptive for Use without a Prescription for All Women of Child-bearing Potential." U.S. Food and Drug Administration. http://www.fda.gov/NewsEvents/Newsroom/PressAnnouncements/ucm358082.htm. Accessed on June 30, 2013.

Ferro, Shaunacy. "Why It's So Hard for Scientists to Study Medical Marijuana." Popsci. http://www.popsci.com/

science/article/2013-04/why-its-so-hard-scientists-study-pot?single-page-view=true. Accessed on June 26, 2013.

"The Fight for Emergency Contraception: Every Second Counts." Center for Reproductive Rights. http://reproductiverights.org/en/emergency-contraception-timeline. Accessed on June 29, 2013.

"Findings and Recommendations of the Personnel Security Board in the Matter of Dr. J. Robert Oppenheimer." The Avalon Project. Yale Law School. Lillian Goldman Law Library. http://avalon.law.yale.edu/20th_century/opp01.asp. Accessed on June 4, 2013.

Fisher, William. "Never Ending Prosecution and Vendetta: The Kafkaesque Story of Sami Al-Arian." Prism. http://prism-magazine.com/2012/05/never-ending-prosecution-and-vendetta-the-kafkaesque-story-of-sami-al-arian/. Accessed on June 4, 2013.

Fletcher, Charles. 2013. *Climate Change: What the Science Tells Us*. Hoboken, NJ: John Wiley and Sons.

Ford, Dan. "How Many Died in Hiroshima?" http://www.warbirdforum.com/hirodead.htm. Accessed on June 3, 2013.

Garwin, Richard L. "Strategic Defense Initiative." http://www.fas.org/rlg/061686_SDI%20DRAFT3.pdf. Accessed on June 16, 2013.

Garwin, Richard L., et al. 1984. *The Fallacy of Star Wars*. New York: Vintage Books.

Gieringer, Dale. 2013a. "The Government's Hundred Years' War on Cannabis." California NORML. http://canorml.org/history/MA_100th_MJ_Anniversary.htm. Accessed on June 24, 2013.

Gieringer, Dale H. 1999; 2013b. "The Origin of California's 1913 Cannabis Law." *Journal of Contemporary Drug Problems*. 26(2): 237–288. Also available online in revised form at http://www.canorml.org/background/ca1913.html. Accessed on June 24, 2013.

Glaser, Charles L. 1984. "Why Even Good Defenses May Be Bad." *International Security*. 9(2): 92–123.

"Global Climate Change." National Aeronautics and Space Administration. http://climate.nasa.gov/index. Accessed on July 1, 2013.

"Global Climate Change and the U.S. Climate Action Report." Hearing before the Committee on Commerce, Science, and Transportation, United States Senate, One Hundred Seventh Congress, Second Session, July 11, 2002. Available online at http://www.gpo.gov/fdsys/pkg/CHRG-107shrg91727/html/CHRG-107shrg91727.htm. Accessed on July 2, 2013.

"Golden Fleece Awards, 1975–1987." Wisconsin Historical Society. http://content.wisconsinhistory.org/cdm/ref/collection/tp/id/70852. Accessed on June 23, 2013.

"Harry J. Anslinger Quotes." Liberty-Tree.ca. http://quotes.liberty-tree.ca/quotes_by/harry+j.+anslinger. Accessed on June 26, 2013.

"Harry J. Anslinger Quotes/Quotations." Uncle Mike's Library. http://www.unclemikesresearch.com/harry-j-anslinger-quotesquotations/. Accessed on June 26, 2013.

Henry, Devin. "Iowa Governor Says Ames Straw Poll Should End, Blames Bachmann's Win." MinnPost. http://www.minnpost.com/dc-dispatches/2012/11/iowa-governor-says-ames-straw-poll-should-end-blames-bachmanns-win. Accessed on June 23, 2013.

Herer, Jack. 2001. *The Emperor Wears No Clothes*, 11th ed. Anaheim, CA: AH HA Publishing, 2001. Also available online at http://www.jackherer.com/thebook/. Accessed on June 26, 2013.

Herken, Gregg. 2002. *Brotherhood of the Bomb: The Tangled Lives and Loyalties of Robert Oppenheimer, Ernest Lawrence, and Edward Teller*. New York: Henry Holt and Company.

Horning, Colleen. 2012. "Magnablend Closes Deal for New Plant Facility at Former SSC Campus." *Waxahachie*

Daily Light. February 1, 2012. Available online at http://
www.waxahachietx.com/news/waxahachie/magnablend-
closes-deal-for-new-plant-facility-at-former-ssc/
article_5c280184-4c71-11e1-be87-0019bb2963f4
.html?TNNoMobile. Accessed on June 10, 2013.

"H.R. 3734." Library of Congress. Thomas. http://
thomas.loc.gov/cgi-bin/query/F?c104:1:./temp/~c104
JRdUDA:e791565:. Accessed on June 27, 2013.

"Illegal Drugs." Gallup. http://www.gallup.com/poll/1657/
illegal-drugs.aspx. Accessed on June 26, 2013.

Inman, Mason. "5 Last-Ditch Schemes to Avert Warming
Disaster." National Geographic News. http://news
.nationalgeographic.com/news/2009/09/090904-
global-warming-fixes-geoengineering.html. Accessed on
June 23, 2013.

*Interference at the EPA: Science and Politics at the U.S. Environ-
mental Protection Agency.* 2008. Cambridge, MA: Union
of Concerned Scientists. Available online at http://www
.ucsusa.org/assets/documents/scientific_integrity/
interference-at-the-epa.pdf. Accessed on July 21, 2013.

"Interviews: Rick Piltz." Frontline. http://www.pbs.org/wgbh/
pages/frontline/hotpolitics/interviews/piltz.html. Accessed
on July 21, 2013.

*In the United States District Court for the Eastern District of
New York.* http://blogs.law.columbia.edu/genderandsexual
itylawblog/files/2010/11/Fifth-Amended-Complaint.pdf.
Accessed on June 29, 2013.

Jones, Steven E. 2006. *Against Technology: From the Luddites
to Neo-Luddism.* New York: Routledge.

J. Robert Oppenheimer "Now I am become death . . ."
Atomic Archive.com. http://www.atomicarchive.com/
Movies/Movie8.shtml. Accessed on June 3, 2013.

Jungk, Robert. 1958. *Brighter Than a Thousand Suns: The
Story of the Men Who Made the Bomb.* Trans. by James
Cleugh. New York: Harcourt, Brace, & World.

Kevles, Daniel J. 2003. "The Strange Case of Robert Oppenheimer." The New York Review of Books. December 4, 2003, 50, 19. http://www.nybooks.com/issues/2003/dec/04/. Accessed on June 4, 2013.

"Key U.S. Missile Interceptor Test Fails, Pentagon Says." Reuters. http://www.reuters.com/article/2013/07/06/usa-military-missile-test-idUSL2N0FC01220130706. Accessed on July 16, 2013.

Koebler, Jason. "Abstinence-only Education Debate Resurfaces." U.S. News. http://www.usnews.com/education/blogs/high-school-notes/2011/12/28/abstinence-only-education-debate-resurfaces. Accessed on June 27, 2013.

"The La Guardia Report—Sociological Study." Shafer Library of Drug Policy. http://www.druglibrary.net/schaffer/Library/studies/lag/concl.htm. Accessed on June 26, 2013.

Lee, Martin A. 2012. *Smoke Signals: A Social History of Marijuana: Medical, Recreational, and Scientific.* New York: Scribner.

Lee, Wen Ho, and Helen Zia. 2001. *My Country Versus Me: The First-hand Account by the Los Alamos Scientist Who Was Falsely Accused of Being a Spy.* New York: Hyperion.

Marbach, William D. 1987. "When Protons—and Politics—Collide." *Newsweek.* June 22, 1987.

McCabe, John. 2011. *Marijuana & Hemp: History, Uses, Laws and Controversy.* Santa Monica, CA: Carmania Books.

McGarvey, Ayelish. 2005. "Dr. Hager's Family Values." *The Nation.* May 30, 2005. Also available online at http://www.thenation.com/article/dr-hagers-family-values#. Accessed on June 29, 2013.

Mckie, Robin. "A True Giant among Men." The Guardian/The Observer. http://www.guardian.co.uk/books/2008/jan/20/biography.features. Accessed on June 4, 2013.

"Medical Marijuana." ProCon.org. http://medicalmarijuana.procon.org/view.resource.php?resourceID=000881. Accessed on June 26, 2013.

Mittelstadt, Michelle. 1993. "Congress Officially Kills Super Collider Project." Lewiston (ME) *Sun Journal.* October 22, 1993. Available online at http://news.google.com/newspa pers?id=kGAgAAAAIBAJ&sjid=umUFAAAAIBAJ&dq= cost%20overrun%20ssc&pg=3808%2C4981568. Accessed on June 10, 2013.

"NASA Langley Research Center's Contributions to the Apollo Program." Langley Research Center. http://www .nasa.gov/centers/langley/news/factsheets/Apollo.html. Accessed on June 9, 2013.

"National Missile Defense." Federation of American Scientists. http://www.fas.org/spp/starwars/program/nmd/. Accessed on June 14, 2013.

"Nixon Tapes Show Roots of Marijuana Prohibition: Misinformation, Culture Wars and Prejudice." CSDP Research Report. http://www.csdp.org/research/shafernixon.pdf. Accessed on June 27, 2013.

Oreskes, Naomi. 2007. "The Scientific Consensus on Climate Change: How Do We Know We're Not Wrong?" In Joseph F DiMento and Pamela Doughman, eds. *Climate Change: What It Means for Us, Our Children, and Our Grandchildren.* Cambridge, MA: MIT Press.

Perl, Martin L. 1968. "The Superconducting Super Collider Project." Stanford Linear Accelerator Center. http://www .slac.stanford.edu/cgi-wrap/getdoc/slac-pub-3943.pdf. Accessed on June 9, 2013.

Peters, E. Kirsten. 2013. *The Whole Story of Climate: What Science Reveals about the Nature of Endless Change.* Amherst, NY: Prometheus Books.

Pianta, Mario. 1988. *New Technologies across the Atlantic: US Leadership or European Autonomy?* Hemel Hempstead, UK: Harvester and Wheatsheaf. Chapter 4: "The Case of the US Strategic Defence Initiative." Available online at http:// archive.unu.edu/unupress/unupbooks/uu38ne/uu38ne0c .htm. Accessed on June 15, 2013.

Piltz, Rick. "Texas State Officials Give a Preview of Climate Science Censorship to Expect under a Perry Administration." http://www.climatesciencewatch.org/2011/10/13/a-preview-of-climate-science-censorship-to-expect-under-a-perry-administration/. Accessed on June 23, 2013.

Plouffe, William C., Jr., 2011. "1925 Geneva Convention on Opium and Other Drugs." In Mark Kleiman and James Hawdon, eds. *Encyclopedia of Drug Policy: The War on Drugs: Past, Present, and Future*. Thousand Oaks, CA; London: SAGE.

Polenberg, Richard. 2002. *In the Matter of J. Robert Oppenheimer: The Security Clearance Hearing*. Ithaca, NY: Cornell University Press.

Price, Derek de Solla. 1963. *Little Science, Big Science*. New York: Columbia University Press.

"Project Apollo: A Retrospective Analysis." NASA History Program Office. http://history.nasa.gov/Apollomon/Apollo.html. Accessed on June 7, 2013.

Reagan, Ronald. 1990. *An American Life*. New York: Simon and Schuster.

[Reagan, Ronald]. "Address to the Nation on National Security." http://www.fas.org/spp/starwars/offdocs/rrspch.htm. Accessed on May 15, 2013.

"The Report of the National Commission on Marihuana and Drug Abuse." Schaffer Library of Drug Policy. http://www.druglibrary.org/schaffer/library/studies/nc/ncmenu.htm. Accessed on June 27, 2013.

Revkin, Andrew C. 2013a. "Bush Aide Softened Greenhouse Gas Links to Global Warming." *New York Times*. June 8, 2005. http://www.nytimes.com/2005/06/08/politics/08climate.html?ei=5090&en=22149dc70c0731d8&ex=1275883200&partner=rssuserland&emc=rss&_r=0. Accessed on July 21, 2013.

Revkin, Andrew C. 2013b. "Editor of Climate Reports Resigns." *New York Times*, June 10, 2005. http://www

.nytimes.com/2005/06/10/politics/11cooney.long.html. Accessed on July 21, 2005.

"RPTS Scott." Executive Session. Committee on Oversight and Government Reform. U.S. House of Representatives. Washington, DC. Interview of: Philip Cooney. Monday, March 12, 2007. Washington, DC. http://oversight-archive .waxman.house.gov/documents/20070319150512-38472 .pdf. Accessed on July 21, 2013.

Rubin, Alissa J., and Holly Idelson. 1991. "Senators Come Out in Favor of Funding Super Collider." *Congressional Quarterly*. July 13, 1991.

Sagan, Carl. 1996. *The Demon-Haunted World: Science as a Candle in the Dark*. New York: Random House.

Sale, Kirkpatrick. "America's New Luddites." Le Monde Diplomatique. http://mondediplo.com/1997/02/20luddites. Accessed on June 16, 2013.

Saul, Rebekah. "Whatever Happened to the Adolescent Family Life Act?" The Guttmacher Report on Public Policy. 1(2) (April 1988). Available online at http://www .guttmacher.org/pubs/tgr/01/2/gr010205.html. Accessed on June 27, 2013.

Scannell, Kate. "Medical Marijuana." SFGate. http://www .sfgate.com/opinion/article/MEDICAL-MARIJUANA-Mr-Attorney-General-Listen-2669856.php. Accessed on June 26, 2013.

Scientific Integrity in Policy Making. 2004. Cambridge, MA: Union of Concerned Scientists. Available online at http:// www.ucsusa.org/assets/documents/scientific_integrity/sci entific_integrity_in_policy_making_july_2004_1.pdf. Accessed on July 2, 2013.

Shulman, Seth. 2004. *Scientific Integrity in Policymaking: An Investigation into the Bush Administration's Misuse of Science*. Cambridge, MA: Union of Concerned Scientists.

"Special Message to the Congress on Urgent National Needs."
The American Presidency Project. http://www.presidency
.ucsb.edu/ws/index.php?pid=8151. Accessed on June 7,
2013.

"Statement by the President." The White House. http://www
.whitehouse.gov/the-press-office/2011/12/08/statement-
president. Accessed on June 30, 2013.

"Statement by U.S. Department of Health and Human Services Secretary Kathleen Sebelius." U.S. Department of
Health & Human Services. http://www.hhs.gov/news/press/
2011pres/12/20111207a.html. Accessed on June 30, 2013.

Stern, Philip M. 1969. *The Oppenheimer Case: Security on
Trial.* New York: Harper & Row.

Stipp, David. 1988. "Federal Plan to Delve into Subatomic
Matter Draws Fire over Costs." *Wall Street Journal.* January
5, 1988.

Stranger-Hall, Kathrin F., and David W. Hall. "Abstinence-
only Education and Teen Pregnancy Rates: Why We Need
Comprehensive Sex Education in the U.S." PLOS-One.
http://www.plosone.org/article/info:doi/10.1371/journal
.pone.0024658. Accessed on June 28, 2013.

"Superconducting Super Collider." rootweb. http://www
.rootsweb.ancestry.com/˜txecm/super_collider.htm#35-41.
Accessed on June 9, 2013.

"Survey of NOAA Fisheries Service Employees." Union of
Concerned Scientists. http://www.ucsusa.org/assets/docu
ments/scientific_integrity/NOAA_Fisheries_Full_Survey_
Results_1.pdf. Accessed on July 21, 2013.

"Taxation of Marihuana." U.S. House of Representatives,
Committee on Ways and Means, Hearings, May 4, 1937.
As cited in "The Marihuana Tax Act of 1937." http://www
.druglibrary.org/schaffer/hemp/taxact/t10a.htm. Accessed
on June 26, 2013.

"Title XX Funding History." Office of Population Affairs. U.S. Department of Health and Human Services. http:// www.hhs.gov/opa/about-opa-and-initiatives/title-xx-afl/ what-is-title-xx/title-xx-funding-history/. Accessed on June 27, 2013.

Trenholm, Christopher, et al. April 2007. *Impacts of Four Title V, Section 510 Abstinence Education Programs: Final Report.* Mathematica Policy Research, Inc. http://www.mathemat ica-mpr.com/publications/PDFs/impactabstinence.pdf. Accessed on June 27, 2013.

"Update on Climate Change Policy." American Geological Institute. Governmental Affairs Program. http://www .agiweb.org/gap/legis107/climate.html. Accessed on July 2, 2013.

United Nations Office on Drugs and Crime. [2010] Policy Analysis and Research Branch. *A Century of International Drug Control.* Vienna: United Nations Office on Drugs and Crime. Available online at http://www.unodc.org/ documents/data-and-analysis/Studies/100_Years_of_Drug_ Control.pdf. Accessed on June 24, 2013.

United States Atomic Energy Commission. 1954. "In the Matter of J. Robert Oppenheimer: Transcript of Hearing before Personnel Security Board, Washington, D.C., April 12, 1954, through May 6, 1954." Washington, DC: Government Printing Office. http://archive.org/stream/united statesatom007206mbp/unitedstatesatom007206mbp_djvu .txt. Accessed on June 4, 2013.

United States House of Representatives. Committee on Government Reform—Minority Staff. Special Investigations Division. "The Content of Federally Funded Abstinence-only Education Programs." http://www.apha.org/apha/ PDFs/HIV/The_Waxman_Report.pdf. Accessed on June 27, 2013.

"US Ballistic Missile Defense Timeline: 1945–2008." Union of Concerned Scientists. http://www.ucsusa.org/

nuclear_weapons_and_global_security/missile_defense/
policy_issues/us-ballistic-missile-defense.html. Accessed on
June 16, 2013.

U.S. Congress, Office of Technology Assessment. 1985. *Ballistic Missile Defense Technologies*. OTA-ISC-254. Washington,
DC: U.S. Government Printing Office. September 1985.

U.S. Congress, Office of Technology Assessment. 1988. *SDI:
Technology, Survivability, and Software*. OTA-ISC-353.
Washington, DC: U.S. Government Printing Office. May
1988.

"U.S. District Court, Denver, Colorado Imposes First Federal
Marihuana Law Penalties." Uncle Mike's Library. http://
www.unclemikesresearch.com/u-s-district-court-denver-
colorado-imposes-first-federal-marihuana-law-penalties/.
Accessed on June 26, 2013.

"U.S. Fish & Wildlife Service Survey Summary. February
2005." Union of Concerned Scientists. http://www.ucsusa
.org/assets/documents/scientific_integrity/fws_survey_sum
mary_1.pdf. Accessed on July 21, 2013.

"Voices of Scientists at FDA: Protecting Public Health Depends on Independent Science." Union of Concerned
Scientists. http://www.ucsusa.org/assets/documents/scien
tific_integrity/fda-survey-brochure.pdf. Accessed on July
21, 2013.

Wales, T. C. "Politics, Personalities and Fear: A New Take on
the Oppenheimer Security Hearings." H-Net Reviews in
the Humanities and Social Sciences. http://www.h-net.org/
reviews/showrev.php?id=6621. Accessed on June 4, 2013.

"What the Research Says . . . Abstinence-Only-Until-Marriage Programs." SIECUS. http://www.siecus.org/_data/
global/images/What%20the%20Research%20Says-Ab-
Only-1.pdf. Accessed on July 27, 2013.

Whitebread, Charles. "The History of the Non-Medical Use
of Drugs in the United States: A Speech to the California
Judges Association 1995 Annual Conference." http://www

.druglibrary.org/olsen/DPF/whitebread.html#TOC. Accessed on June 26, 2013.

Willard, E. Payson. 1994. "The Demise of the Superconducting Super Collider: Strong Politics or Weak Management." *PMI Canada Proceedings*. 1994: 1–7. Available online at http://iems.ucf.edu/admissions/graduate/exams/Sp05-EngineeringManagement1.3.pdf. Accessed on June 10, 2013.

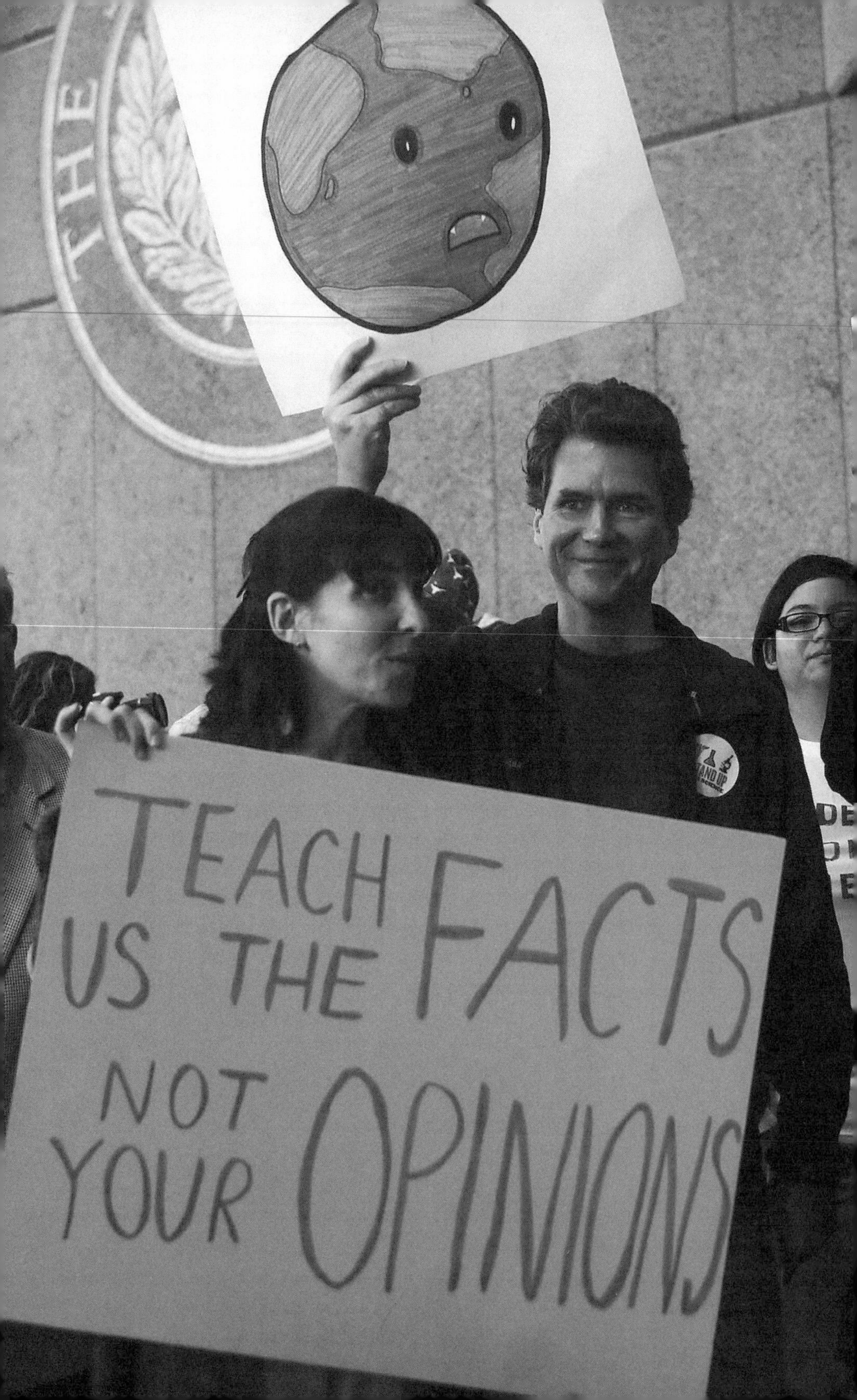
THE
TEACH US THE FACTS NOT YOUR OPINIONS

Introduction

The topic of the relationship between science and politics is one of interest to a great variety of individuals. This chapter provides an opportunity for a range of writers to express their views on some aspects of this topic.

THE POLITICS OF HUMAN EMBRYONIC STEM CELLS

Sandy Becker

When people seriously disagree about issues, they often resort to politics. (It's better than resorting to violence, right?) At this point people seriously disagree about whether it's acceptable to make and use human embryonic stem cells (ES cells). So, making and using human ES cells have become a political issue. Some people want it banned or at least restricted; some don't.

ES cells can be chemically teased and tickled by researchers to become all the different cell types in the body, such as nerve cells, muscle cells, blood cells, or brain cells. So if someone has

Science writers display signs during a news conference where a new science curriculum for Texas public schools was discussed in 2008. The State Board of Education prepared to propose new standards that would encourage middle school students to discuss alternative explanations for evolution. (AP Photo/Harry Cabluck) lost or damaged a certain group of cells due to injury or disease, he or she could receive a transplant to replace them. Of course, the science of transplanting cells into human subjects is still in its infancy; scientists can't really do this yet. But if they had the cells, maybe they could. Meanwhile, they can use the stem cells and their differentiated descendants as models to study disease processes.

The key words that people seriously disagree about here are "human" and "embryonic." There are ES cells made from the embryos of animals, usually mice, and hardly anybody objects to making them and using them for research. There are human stem cells that are made from adult body cells, not from embryos, and hardly anybody objects to making or using those cells either. But making ES cells from a human embryo deprives the embryo of the opportunity to develop into a baby, and some people object very much to that.

ES cell lines are made from very early embryos, consisting of a few dozen, up to a 100, cells. At this point the embryo is called a blastocyst. Blastocysts of humans and other mammals are tiny blobs of cells, smaller than a pinhead. Human embryos that can be used to make stem cell lines are grown from eggs fertilized in a petri dish, just as they are for in vitro fertilization (test tube babies). If they are used for in vitro fertilization (IVF), they will be implanted into the mother's uterus at the blastocyst stage. This isn't a very efficient procedure. If two or three blastocysts are implanted, there is perhaps a 40 percent chance of a successful pregnancy. Embryos not used immediately can be frozen indefinitely in liquid nitrogen for future use.

If the blastocyst is used to make ES cells, it will be cultured in a plastic dish on a layer of "feeder cells," and with luck, some of the cells will become ES cells. This is also a very inefficient process—ES cell lines will be successfully established less than 50 percent of the time.

Mouse ES cells have been made and used for research since 1981. A great deal has been learned from them about embryonic development and about many disease processes. But mice are quite different from people, and there is only just so much that can be learned from studying them. Besides, we can't go transplanting mouse cells into humans to replace tissues they've lost or damaged! So in 1998 the first human ES cells were made, and that's when the political debate began. Immediately, some people were outraged that human embryos had been used in this way. To these folks, there is simply no excuse for "destroying" a human embryo. To others, the potential medical benefits outweigh this concern.

The Dickey–Wicker Amendment passed by Congress in 1995 stated that scientific research grants from the federal government, usually the National Institutes of Health or National Science Foundation, could not be used for research using human embryos. (The research establishing the first human ES cell lines in 1998 was supported by the University of Wisconsin and Geron Corporation.) Embryo research was not banned; it just couldn't be done with taxpayers' money. Making and studying human ES cells has *never* been banned in the United States; it just couldn't be done with taxpayers' money. In 2001, President George W. Bush implemented a policy under which a few specified human ES lines could be used for research with federal grant money. In 2009, President Obama expanded the list of cell lines acceptable for federal funding. A key issue in allowing cell lines to be eligible for federal funding has been whether the donors of the egg and sperm used to make the embryos had given their "informed consent" for them to be used for research. Only cell lines in which this consent can be clearly established are eligible for federal funds.

Meanwhile, several state legislatures took up the slack. California, for example, allocated $3 billion for a 10-year program in support of ES cell research. The state of Connecticut allocated $100 million over 10 years for the Connecticut Stem Cell Initiative. The lab I work in has been funded for several years by this money. Since I do research using human ES cells, I guess it's clear I'm in the group that feels the likely scientific advances now and medical benefits in the future outweigh my concern about depriving an embryo of the chance to grow into a baby.

People who disagree feel that even a very early embryo is a living person and has the right to continue developing into a baby. I would remind them that even an embryo conceived in the old-fashioned, low-tech way, by two people having sex, has less than a 50–50 chance of surviving even long enough so the mother knows she is pregnant. There is a lot of failure in human pregnancy. I would also remind them the alternatives: "extra" embryos left over after IVF will be frozen in liquid nitrogen as long as someone, usually the donors, pays for their storage. Then they will be discarded. Is this a better fate than becoming an ES cell line, able to divide forever and differentiate into many different cell types?

Sandy Becker has been doing stem cell research for over 25 years at Wesleyan University in Middletown, CT. She moonlights as a science writer, and has recently begun writing children's books for her four-year-old grandson.

KITZMILLER V. DOVER IN FOCUS
Glenn Branch

Taking place in a federal court over 40 days in 2005, *Kitzmiller v. Dover* was the trial in which the unconstitutionality of teaching intelligent design in the public schools was established. Tammy Kitzmiller was the mother of two children who attended Dover High School, in Dover, Pennsylvania, and she was one of 11 parents who sued the Dover Area School Board over a policy about the teaching of evolution adopted in October 2004. The policy read, in part, "Students will be made aware of gaps/ problems in Darwin's Theory and of other theories of evolution including, but not limited to, intelligent design. Note: Origins of life will not be taught."

The board adopted the policy after a summer of wrangling over what biology textbooks to use. The school's teachers and the district's administrators wanted to adopt a standard textbook. But the chair of the board, William Buckingham, complained that the textbook was "laced with Darwinism" and vowed to seek a textbook in which both evolution and creationism were presented. Later, Buckingham tried to have the board adopt *Of Pandas and People*, a supplementary textbook advancing intelligent design. Although he was unsuccessful, in early October 2004, 60 copies were anonymously donated to the district as "reference material" for classrooms, which seemed to represent a compromise.

But on October 18, 2004, the board formulated and voted to adopt the policy, which, as the first district policy requiring the teaching of intelligent design, swiftly became a matter of national controversy. To implement the policy, the board approved a statement for teachers to read to students, specifically recommending that students read *Of Pandas and People*. In December 2004, the 11 parents—represented by Americans United for Separation of Church and State, the American Civil Liberties Union of Pennsylvania, and a private law firm, Pepper Hamilton LLP—filed a federal lawsuit against the board, arguing that the policy violated the Establishment Clause of the First Amendment.

Representing the board was the Thomas More Law Center, a religious nonprofit law firm that had been seeking a test case on intelligent design. Although the board was relying on the published advice of the Discovery Institute, the de facto institutional headquarters of intelligent design was in fact unhappy about the case, calling the board's policy misguided and recommending its replacement by a policy requiring "full disclosure of the scientific evidence for and against Darwin's theory." Also critical of the policy, but for a quite different reason, were the Dover Area High School science teachers, who in January 2005 announced their collective refusal to read the required statement.

The attorneys for the plaintiffs (the parents) and the defendants (the school board) spent the spring and summer of 2005 preparing for the trial. The legal team for the defendants was in a state of disarray after a dispute between the Thomas More Law Center and the Discovery Institute caused three expert witnesses associated with the Discovery Institute who were expected to testify on behalf of the defendants to withdraw from the case. Later, the Foundation for Thought and Ethics, the publisher of *Of Pandas and People*, sought to intervene by becoming a codefendant in the case. If successful, it reportedly would have called those expert witnesses, but it was not permitted to intervene.

The trial began on September 26, 2005. Presiding was Judge John E. Jones III, a churchgoing Republican appointed to the bench by President George W. Bush. The basic issue before the court was whether the policy violated the Establishment Clause of the First Amendment. Did the policy endorse a religious view? The testimony thus centered on questions such as: Is intelligent design (and the idea that there are "gaps/problems" in evolution) a religious view? What was the board's primary purpose in adopting the policy? And because the board's defense was that the policy was intended to promote science education, there was a further question: Is intelligent design science?

The plaintiffs argued that intelligent design's criticisms of evolution were scientifically unwarranted, calling a cell biologist, Kenneth R. Miller, and a paleontologist, Kevin Padian, to the stand to explain the evidence for evolution. Michael J. Behe, a biochemist testifying for the defendants, was forced to concede the absence of any "detailed rigorous accounts of how intelligent design of any biological system occurred" from the scientific literature. The plaintiffs further argued in detail that intelligent design was a development of previous forms of creationism, citing the fact that references to "creation" in drafts of *Of Pandas and People* were replaced with references to "design" in the published book.

As for the board's purpose in adopting the policy, members of the Dover community testified to a campaign on the part of Buckingham and his ally Alan Bonsell to promote their own religious views. Buckingham argued against adopting the textbook "laced with Darwinism" by saying, "Two thousand years ago someone died on a cross. Can't someone take a stand for him?" Bonsell proved to have been involved in the *Of Pandas and People* donation and, apparently, to have lied about it. After the trial concluded but before the verdict was announced, the November 2005 election took place. Turnout was high, and all of the incumbent school board members running for reelection were defeated.

On December 20, 2005, the decision in *Kitzmiller v. Dover* was issued, and the plaintiffs won their case. In his decision, Judge Jones concluded that it was "abundantly clear that the Board's [intelligent design] Policy violates the Establishment Clause. In making this determination, we have addressed the seminal question of whether [intelligent design] is science. We have concluded that it is not, and moreover that [intelligent design] cannot uncouple itself from its creationist, and thus religious, antecedents." The board was held responsible for $1,000,011 for the legal fees and damages of the plaintiffs, with the $11 divided equally among the 11 parents.

What was the effect of the decision? Locally, it helped to reconcile a polarized community; the new board voted in January 2006 not to appeal the decision. Elsewhere, although the decision is binding only in the Middle Federal District of Pennsylvania, its influence was widely felt. Legislators in Indiana and Utah who had promised in 2005 to introduce bills requiring the teaching of intelligent design revised their plans in 2006. In general, in the wake of *Kitzmiller v. Dover*, attacks on evolution in public education are increasingly avoiding calling for creation science and intelligent design to be taught, instead favoring the strategy of belittling evolution (as controversial or just a theory).

Further Information

Humes, Edward. 2007. *Monkey Girl: Evolution, Education, Religion, and the Battle for America's Soul*. New York: Ecco.

Judgment Day: Intelligent Design on Trial [film]. 2007. Directed by Gary Johnstone and Joseph McMaster. Boston: NOVA and Vulcan Productions.

Lebo, Lauri. 2007. *The Devil in Dover: An Insider's Story of Dogma v. Darwin in Small-Town America*. New York: The New Press.

Miller, Kenneth R. 2008. *Only a Theory: Evolution and the Battle for America's Soul*. New York: Viking.

Tammy Kitzmiller et al. v. Dover Area School District et al., 400 F. Supp. 2d707 (M.D. Pa. 2005)

Glenn Branch is deputy director of the National Center for Science Education (ncse.com), a nonprofit organization that defends the teaching of evolution and climate science.

SCIENCE IS NOT A LIBERAL CONSPIRACY
Allan B. Cobb

Scientists are by their very nature skeptical. Skepticism is one of the mechanisms that drives science forward. A scientist observes something and wants to find out why or what caused that particular observation. The scientist looks to the work that has been done in the past and may be skeptical about the explanation. The scientist then uses the scientific method to look at the mechanisms that cause their observation. They may find out that the original explanation is correct or they may find a better explanation. Either way, this occurs because the scientist was skeptical.

When science is brought into a political debate, the people who oppose the general views of the science are also considered skeptics. However, their skepticism is not based on a scientific understanding but rather on a political view. Instead of using the scientific method to further the understanding of the science, they try to obscure and confuse the science to make people believe that the science is wrong. Instead of using science to prove their point, they use methods of propaganda to convince people that the science is wrong. They do this by promoting a general confusion about the science.

This use of skepticism to prove science wrong for political reasons has occurred many times and with many different issues. It happens over issues with human health when compared to pollutants, the linking of cancer to causes such as tobacco use, and effects of industry and chemicals on the environment. Today, one of the most obvious examples is with global warming or climate change.

The science behind climate change is pretty clear. Scientists have a really good idea of what is causing the current changes in our climate. Analysis of the scientific work clearly shows that scientists who work in the field of climate studies are convinced that humans are directly affecting the climate. Even though the science is very clear, there are still areas where more knowledge is needed. Continuing scientific research is slowly filling in these gaps in knowledge. If scientists are convinced that current global warming is occurring due to human influences, why is there even a debate about it?

Whenever science and politics mix and oppose each other, politicians become involved in the debate. Politicians are seldom scientists, yet they make the decisions. Some politicians are not willing to accept the findings of science because they are supported by the businesses and industries that would be hurt by acceptance of the science. In the case of global warming, Republicans promote skepticism about the science to protect the oil industry. How does a political party promote skepticism of science?

Scientists write about science mainly for other scientists. Scientists publish their results in scientific journals that are intended to be read by other scientists who work in the same field. Scientists use a process called peer review. Peer review means that a few other scientists read the papers prior to publication and look for errors in the research. When the papers are published other scientists read them and use their own skepticism to further the understanding. Because these scientific papers are very technical, few people in the general public ever read them. The papers are filled with technical jargon and most people do not have the scientific background to understand. Instead of reading the actual science, the general public learns about the latest science through the media. The media includes newspapers, magazines, TV news, and the Internet.

Unfortunately, many of the people in the media who report on science news are not scientists; so they depend on interviews with scientists to gather information to produce articles for the general public. Sometimes they do not accurately represent the science because scientific concepts may be difficult for the general public to understand, and sometimes it is because of a particular political bias written into the story. When news stories are written with a political bias, they are actually a form of propaganda. As the news becomes more aligned with politics, the problem of getting accurate scientific information increases. Today, with so many news outlets in print, on TV, and on the Internet, it is very easy for people to get their news only from a source with which they share a similar political view.

Currently, one of the hottest scientific—political debates surrounds global warming. Republicans, who generally support big business, are trying to convince their political base that the science is wrong. They claim that liberals are wrong in using science to further their political agenda. It is not unusual to hear Republicans claim that liberals are really just trying to destroy the United States and world economy to promote Communism. They make claims such as that scientists are just interested in getting grant money rather than doing good science. They claim that the climate has changed before and humans are too insignificant to make a difference. They also make claims such as that scientists haven't studied the subject enough to know or that their models are not accurate enough to make predictions. All of these types of claims are meant to confuse the public about the science and promote a general skepticism about the science. They try to make it seem as if the science is part of a big liberal conspiracy. Unfortunately, creating confusion is much easier than promoting education. With all the information available today, how can you be sure what is true and what is propaganda?

The first and most obvious way to tell the difference between science and propaganda is to look at the source of the information. A politician trying to tell you about science is a clue that you should be skeptical about his or her message. You need to look for more reputable sources of information. Look for sources that have information written by scientists. There are many magazines and web sites that promote good scientific information about many different science topics. Read these and educate yourself about a topic. Use your own skepticism to guide your learning. Learn as much as you can about a subject and use that knowledge of the subject to create your own informed opinion.

Allan Cobb is a freelance writer who lives in Central Texas. He specializes in nonfiction topics for young adults and science educational materials.

ROWLAND'S RECIPE FOR CLIMATE TREATY SUCCESS
Joel Grossman

As a sports-crazy youth hitchhiking cross country from Los Angeles to Ohio, visiting Yosemite and Yellowstone National Parks along the way, Frank Sherwood Rowland could easily have turned into a "Dharma Bum" in a Jack Kerouac novel. But a love for math and science applied to atmospheric chemistry sent him down a different road, where he embraced the dual role of scientist and media spokesperson for a great worldwide environmental crusade to limit global greenhouse gas production. One result was the Montreal Protocol, perhaps the most successful greenhouse gas treaty ever implemented, as it dramatically reduced production and use of man-made (synthetic) chlorofluorocarbon (CFC) molecules destroying the stratospheric ozone layer protecting planet Earth from ultraviolet (UV) light wavelengths known to trigger skin cancers.

Rowland's political saga with CFCs and atmospheric chemistry started innocently enough. In the early 1970s, using a newly created scientific instrument, scientist James Lovelock of Gaia theory fame detected totally synthetic molecules (CFCs) in the upper atmosphere for the first time. In the 1930s, CFCs began their metamorphosis from laboratory curiosities into widespread use. CFCs were considered ideal industrial molecules because they were very stable (inert), relatively nontoxic, and incredibly useful as foams, refrigerants, cleaning agents, and aerosol propellants. For Rowland, the fate of CFC molecules in the upper atmosphere was a basic chemistry problem that could be solved using radioactive isotopes (tracers) as part of an annual grant from the U.S. Atomic Energy Commission.

Surprisingly, the same CFC molecules so stable at sea level shed chlorine atoms when under bombardment from intense UV light in the upper atmosphere. More shockingly, Rowland and his colleagues found that the chlorine liberated from CFC molecules was amazingly effective at destroying stratospheric ozone molecules, which form a layer screening life on Earth's surface from UV wavelengths capable of causing cell mutations and skin cancers. With CFC production peaking at just over a million metric tons per year, the research was more than just academic.

"So, beyond the science there were political and environmental questions," Rowland told interviewers a decade after sharing the 1995 Nobel Prize in Chemistry. There was also an "instinctive science fiction aspect" to a story where aerosol propellants in underarm deodorant spray cans were destroying Earth's ozone layer. This almost comic-book sci-fi aspect transformed the science into front page international news, and captured the attention of the political establishment.

Rowland calculated that one of the earliest news stories on his work, published by UPI in September 1974, ran in 400 newspapers and reached 100 million people. This resulted in invitations to talk at universities and to state legislatures. By June 1975, the state of Oregon passed a law banning CFC propellants in aerosol sprays. Soon the ban became nationwide, and the CFC debate became international. The 1985 detection of the Antarctic ozone hole broke down international resistance from Great Britain and France, which had suspected a plot by competitors to derail production of the ozone-destroying

Concorde supersonic transport jet. By 1987 there was an international treaty, the Montreal Protocol, which outlined a path for ending most uses of CFCs and other man-made ozone-depleting chemicals.

Rowland had a leg up on most scientists in the public relations game with the media, having worked on the student newspaper in college. More typically, scientists are subordinated to, muzzled, buffered, or filtered by public relations departments and professional news spinners. Rowland, on the other hand, quickly realized that he could answer his own phone and do a better job communicating directly with TV, magazines, newspapers, and later the Internet digital media. The results speak for themselves: 13 years from research publication to international environmental treaty.

Rowland maintained his scientific credibility by combining his do-it-yourself media outreach with a meticulous policy of sticking to his areas of scientific expertise and understanding. This careful stance created enormous credibility that paid off in the political arena with all sides taking Rowland very seriously. Even international treaty opponents like scientific gadfly Fred Singer, who questioned the significance of the ozone hole and greenhouse gases, conceded the correctness of Rowland's science. In the political realm, this credibility translated into over 190 nations signing onto and implementing the Montreal Protocol protecting the stratospheric ozone layer.

Realizing that having "Nobel-Prize-winner" associated with his name changes the public perception, Rowland was very careful to keep his name off of petitions, causes and campaigns outside his scientific expertise and understanding. In other words, Rowland understood that his credibility and ability to influence events and political legislation depended on carefully controlling and limiting how his name and endorsements were used.

Since greenhouse gases are also associated with global warming, this meant walking a delicate tightrope. But in 2001, when the newly elected administration of President George W. Bush asked the National Academy of Sciences (NAS) to form a committee to educate the cabinet on global warming, Rowland agreed to serve. The cabinet was divided on the issue, but President Bush recognized the reality of global warming in a Rose Garden speech. However, after 9/11 the focus shifted to homeland security and terrorism, and, said Rowland, global warming dropped off the presidential radar screen.

Rowland acknowledged the value of a precautionary approach to global warming, but refused to endorse any proposed solutions because, "the knowledge of how the global climate system works is insufficient." Indeed, stratospheric air samples examined for CFCs contained over 200 different molecules, many of which, like methane from rice paddies and cattle farming, are natural and interact in ways affecting diverse variables like cloud cover that impact global warming in complex ways still being deciphered. The Bush administration learned from their weekly NAS seminars that the laudable goal of removing the pollutant sulfur dioxide from burning coal resulted in increased global warming via reducing the atmospheric sulfate layer. Ever the cautious experimental chemist, Rowland was wary of "experimenting" on complex problems like global warming with insufficient knowledge, as even with the best of intentions it might be a recipe for creating new global climate problems.

Joel Grossman is a Southern California–based independent writer and book editor, writes the Internet-based Biocontrol Beat blog (WordPress). He also reports on the Entomological Society of America for a print newsletter, The IPM Practitioner. *Mr. Grossman helped the U.S. EPA and United Nations develop a research agenda on alternatives to the ozone-depleting agricultural and industrial fumigant, methyl bromide, under the Montreal Protocol.*

THE EVOLVING DISPUTE
OVER TEACHING DARWINISM
Phill Jones

In 1859, Charles Darwin published *On the Origin of Species.* In Darwin's view, life on Earth changes through evolution, a process fueled by random variations of physical traits and driven by a pitiless survival of the fittest selection mechanism. Many Christians fiercely rejected Darwinism; they believed in an amiable universe that reflected intelligent design by a loving Creator. The early 20th century saw reconciliation between some religious leaders and proponents of Darwinism. Theistic evolution, for example, emerged as a philosophy that evolution served God's purpose. Some resisted efforts to reconcile faith and religion. In the United States, the Protestant fundamentalist movement opposed compromise.

Following the brutal World War I, the flamboyant, rebellious Jazz Age culture blossomed in 1920s America. Many saw a declining morality and blamed Darwinism for influencing people to replace religious faith with materialism. Opponents of Darwinism also condemned the use of the natural selection concept to justify merciless business practices and the American eugenics movement, which called for the imprisonment or sterilization of those deemed unfit for the struggle of existence.

Fears arose that Darwinism would erode students' faith. By now, textbook authors had replaced a theistic version of evolution with a Darwinian view of human evolution, a view supported by fossil discoveries. A record attendance in American high schools—from about 200,000 during the late 19th century to about 2 million by 1920—created a sense of urgency about terminating discussion of Darwinism in the classroom.

An American crusade against teaching evolution gained momentum. William Jennings Bryan, a leader in the fundamentalist movement and the antievolutionist crusade, insisted that public school teachers must teach what taxpayers desire to have taught. And that did not include evolution. About 20 state legislatures considered bills to ban the teaching of evolution in public schools. Some became law, such as Tennessee's Butler Bill of 1925. The law created a misdemeanor punishable by fine for any Tennessee public school teacher "to teach any theory that denies the story of the Divine Creation of man as taught in the Bible, and to teach instead that man has descended from a lower order of animals" (Tennessee Anti-Evolution Act, 1925).

The New York City–based American Civil Liberties Union (ACLU) saw the antievolution statute as a threat to academic freedom. The organization advertised for a Tennessee teacher willing to test the law in court. Substitute biology instructor John T. Scopes accepted that role. The defense team included the renowned attorney Clarence Darrow; the equally famous William Jennings Bryan joined the prosecution's team.

The defense wanted to show that the antievolution statute violated state and federal constitutions. The judge, however, rejected the defense's objections to the statute. The jury returned a guilty verdict. Anticipating that they would lose an appeal of the verdict in state court, ACLU leaders planned to bring the *Scopes* case to the U.S. Supreme Court. But the Tennessee Supreme Court overturned Scopes' conviction on a technicality, saying that, "We see nothing to be gained by prolonging the life of this bizarre case" (*Scopes v. State*, 1927).

Antievolution laws persisted for decades. Authors omitted Darwinism from new textbooks, and teachers avoided the subject.

A Satellite Launches More Controversy

In 1957, the world heard a "beep" from space. As it orbited the Earth, the Russian satellite Sputnik transmitted the sound of victory. The fact that the Soviet Union had beaten the United States into space seemed to highlight the second-rate standing of American science education. The U.S. government poured more than a billion dollars into efforts to update and expand science education programs. With the change in science curriculum, new textbooks included Darwin's theories, now bolstered by studies in paleontology and population genetics. The reappearance of Darwinism in the classroom rekindled an old controversy.

For the academic year 1965–1966, administrators of a high school in Little Rock, Arkansas, adopted a textbook that contained a chapter on the evolution of humans. Biology teacher Susan Epperson faced a dilemma: The state of Arkansas banned the teaching of evolution. Should she use the new book and face criminal prosecution? She filed a case in a state court, asking the court to declare the 1928 Arkansas antievolution statute as unconstitutional. The court agreed that the "monkey law" was invalid. The state of Arkansas appealed the decision. Unlike the *Scopes* case, *Epperson v. Arkansas* reached the U.S. Supreme Court.

In 1968, the Supreme Court struck down Arkansas' anti-evolution law. "The law's effort," the Court wrote, "was . . . an attempt to blot out a particular theory because of its supposed conflict with the Biblical account, literally read." The law violated the First Amendment doctrine of church and state separation. The government "may not be hostile to any religion or to the advocacy of no-religion" (*Epperson v. Arkansas*, 1968).

Antievolutionists Struggle to Adapt Tactics

During the 1970s, antievolutionists tried a different approach: They depicted creationism as "creation science" and demanded equal time in the classroom for their alternative to evolution. By the 1980s, legislators in about 27 states had introduced bills requiring equal time for creation science. Several were signed into law. The U.S. Supreme Court struck down a Louisiana equal time law in *Edwards v. Aguillard* (1987). The Court said that the law violated the First Amendment by using government support to achieve a religious purpose.

Antievolutionists responded to the *Edwards* decision by promoting "intelligent design," the idea that an Intelligent Designer must be responsible for at least some traits of life on Earth. When Pennsylvania's Dover Area School District required teachers to read a statement about intelligent design to students, a group of parents sued, alleging that the policy was unconstitutional. In *Kitzmiller v. Dover Area School District* (2005), the judge decided that the policy was an attempt to impose a religious view of biological origins, and this violated the U.S. Constitution's command to separate church and state.

State legislators continue to try new antievolutionist strategies. Many current state bills and laws encourage teachers to discuss faults of the "controversial" theory of evolution. Critics argue that the laws provide a backdoor for teachers to introduce religious beliefs into science curriculum. If one state's new breed of monkey law eventually enjoys success, then it will undoubtedly be copied by other state legislatures. That's a type of natural selection.

References

Epperson v. Arkansas, 393 U.S. 97, 104, 109 (1968).

Scopes v. State, 289 S.W. 363, 367 (Tenn. 1927).

Tennessee Anti-Evolution Act, Tenn. Code Ann. §49—1922 (repealed 1967).

Phill Jones earned a PhD in physiology/pharmacology and a JD. He worked 10 years as a patent attorney, specializing in biological, chemical, and medical inventions. As a freelance writer, he writes articles and books in the areas of general science, forensic science, medicine, history, law, and business.

THE POLITICS OF GENETICALLY MODIFIED FOODS
Bill Loftus

Biotechnology applied to crops sometimes generates extensive controversy. Even golden rice, a crop scientists say they bred specifically to benefit people suffering devastating diseases in impoverished areas, has drawn widespread protest. Most of the protests result from the use of genetic engineering, specifically inserting a gene from one species to another. In the case of golden rice, two genes, one from a bacterium and the other from corn, were transferred into rice. The rice acquired a new trait, the ability to create and store beta-carotene in the endosperm of its seeds, as a result.

Proponents of golden rice said creating rice capable of supplying dietary beta-carotene, also known as pro-vitamin A, could reduce disease and death among impoverished people without access to fresh fruits and vegetables. Despite extensive media coverage of golden rice as a potential breakthrough to benefit human health, protests mounted. Nearly a decade after its debut, golden rice drew activists in the Philippines to destroy a field of golden rice plants before they could be harvested. Media reports differed over descriptions of whether farmers or activists opposed to genetically modified organisms (GMOs) led the destruction.

Courtroom conflicts over GMOs have been persistent over the last two decades, focused mostly in the courtrooms. In southern Idaho, a legal fight focused on sugar beets that had been genetically engineered to resist herbicides. Opponents of the Roundup Ready™ technology developed by Monsanto waged a legal campaign to prevent GMO sugar beets from being grown. Proponents argued that the technology posed no health threat, but that it would allow growers to survive economically. In 2012, federal regulators gave final approval to use of the Roundup Ready™ sugar beets by farmers.

The legal battles over the use of genetic engineering to create GMOs boil down to whether adding genes from other species, or rearranging them within the same organism, can create health threats to people or animals, or environmental threats. GMO opponents have seized on studies that reported corn pollen containing bacterial genes to control insect pests also hurt monarch butterflies. GMO proponents have seized on other scientific reports and regulatory agency studies that found no nutritional differences between GMO plants and those produced with conventional plant breeding methods.

In 1994, the excitement over biotechnology was focused more on its novelty than on whether it was a threat or a promise to agricultural producers. In that year, a group of genetic engineers in California introduced a tomato they said would survive shipping and taste good. American consumers who tried the tomatoes in supermarkets shortly after the new fruit's introduction agreed. The Flavr Savr® tomato's genetic advance relied on rearranging tomato genes, specifically the gene that controlled softening, and in turn flavor.

The new tomato passed food safety reviews by the Food and Drug Administration and appeared in California grocery stores. Although early opponents of genetic engineering raised objections based largely on a lack of sufficiently detailed studies, shoppers liked the new tomato. The scientists who created the tomato found, however, that they had much to learn about the produce business. Popularity among shoppers could not overcome issues like shipping and the difficulties of getting high quality produce to the stores. Calgene sold its technology to Monsanto, which ended production of Flavr Savr® tomatoes.

The lessons of the Flavr Savr® tomato continue to influence thinking about biotechnology and food. Economists and scientists say the issues that affect introduction of the technology range from food safety to its economic effects. In the case of food safety, uncertainty is one of the key issues raised by opponents. The opposing views are often divided into two main areas: consumer-oriented traits and grower-oriented traits. Consumer-oriented traits refer to those like the Flavr Savr® tomato and its genetic adjustments that lead to better-tasting tomatoes. Producer-oriented traits refer to those that help farmers achieve higher crop yields or use less pesticides to protect their crops from insects or weeds.

Those who oppose biotechnology's use in food crops argue it is important to know as much about whether genetic transformation of a food could create allergic reactions in those who eat the food or more subtle changes in the body's responses that could damage health years later. Proponents say the food safety issue is much less mysterious. Genetically modified crops approved for human use are the same nutritionally as any other food.

Proponents of producer-oriented traits argue that farmers gain significant economic advantage through the use of genetic modifications, such as those that make crops resistant to herbicides. Weed control in sugar beet fields must rely either on workers hand-hoeing rows, a physically hard and financially costly method, or by chemical controls, costly but less labor intensive. Chemical weed control requires a sugar beet that can survive sprays that kill weeds. Opponents argue that the increased use of herbicides can damage soil and water and those environmental effects outweigh the apparent savings to farmers.

For many crops, farmers have already decided. Most of the corn and soybeans grown in the United States and in some other nations are already genetically modified to be herbicide resistant. In some parts of the world, notably Europe, consumers resist buying GMO foods, including animals that have been fed them. Japan and Korea both briefly shut markets to U.S. wheat in 2013 after the discovery of herbicide-resistant wheat in an Oregon field. Extensive testing and government investigations showed that the discovery was limited to that particular field and that genetically modified wheat had not entered commerce, convincing both nations to begin buying U.S. wheat again.

The next form of biotech foods that may reach Americans is French fries. An industry giant is asking U.S. regulators to approve new spuds that produce less acrylamide, a chemical found in cigarettes and deep-fried potatoes, among other foods. Studies with rodents suggest acrylamide is capable of causing cancer. The new potatoes, like the Flavr Savr® tomato, contain only potato genes that have been rearranged. The question for consumers may become do you want fries with acrylamide, or do you want biotech fries without acrylamide?

Bill Loftus is a science writer based in Genesee, Idaho.

SCIENCE, POLITICS, AND HIGH ENERGY PHYSICS
Michael Perricone

Can we afford to learn what the universe is made of? Do we have the money to find out how it started and how it works?

What's the payoff in solving the mysteries of the ghostly neutrinos, invisible dark matter, and missing antimatter?

High energy physics is the science looking for the answers to these questions. It uses the biggest and most complicated machines ever built, particle accelerators and detectors, in the quest to answer the biggest and most basic questions about our existence.

And that poses the biggest question about high energy physics itself: can the U.S. government afford to keep funding these massive, complex machines with the single goal of answering our deepest questions?

Particle physics is a prime example of fundamental science research: that is, research aimed solely at producing knowledge without direct practical applications. As part of its funding of basic research, the U.S. Department of Energy's Office of Science operates major accelerator projects at Fermilab and Argonne labs in Illinois, Brookhaven Lab in New York, Thomas Jefferson Lab in Virginia, Oak Ridge Lab in Tennessee, and Stanford Accelerator Center in California. But the story of Fermilab's Tevatron accelerator is a unique saga.

For nearly 30 years, the United States was home to the "energy frontier" of the particle physics world. On the Illinois plains 30 miles west of Chicago, an underground particle accelerator created millions of high energy collisions each second between protons and antiprotons. Antiprotons are real-life antimatter, the stuff straight out of *Star Trek*, custom-made at Fermilab.

At Fermilab (Fermi National Accelerator Laboratory), the only U.S. lab focused solely on high energy physics, U.S. scientists and visitors from around the world sent these subatomic particles and antiparticles whizzing through the Tevatron's four-mile underground ring faster than 99.99 percent of the speed of light. They crashed the two particle beams together in the middle of two 5,000-ton particle detectors, each about the size of a small apartment building, to see what came flying out. The collision energy was the highest produced in any collider in the world. Billions of times every day, Fermilab demonstrated Einstein's famous equation: $E = mc^2$.

Science at the Tevatron was about as cool as science gets—literally. The Tevatron's multiton magnets were cooled with liquid nitrogen and liquid helium to 4.5 K—4°C above absolute zero (–450°F). At these "superconducting" temperatures, electrical resistance is close to zero, meaning almost no energy is lost to the heat of resistance, no matter how much electricity flows through the cables.

The Tevatron was famed for its 1995 discovery of the top quark, the short-lived and most massive of the subatomic particles. In the earliest days of the universe, the top quark was freewheeling along with other particles. Tevatron collisions were aimed at getting close to that early universe environment.

By the end of the Tevatron's 28-year run in late 2011, Fermilab's annual budget was $408 million. At its peak in the Tevatron years, Fermilab had more than 2,000 full-time employees, including hundreds of scientists, engineers, and technicians—all pumping their salaries right back into the economy.

In addition to its top quark legacy, the Tevatron and its particle detectors provided the research environment to produce more than 1,000 PhD degrees. And thousands of elementary-to—high school students have been touched by Fermilab's educational and outreach programs over the years. Some of them might grow up to be scientists themselves.

But the payoffs in high energy physics have been practical as well as intangible. They have affected lives directly as well as feeding our hunger for knowledge.

If you have ever had an MRI scan, you have experienced medical technology directly descended from particle physics. Superconductivity, which was first used on an industrial scale in 1,024 Tevatron magnets, creates the intense magnetic fields used in magnetic resonance imaging.

Particle beams such as protons and neutrons treat millions of cancer patients each year. Particle accelerators can be found at every major medical center in the United States. And particle beams can scan cargo containers without opening them, an important security contribution.

Particle physicist Tim Berners-Lee of CERN, the multinational European particle physics lab in Switzerland, developed the online tool that became the World Wide Web in 1989, so that he and other particle physicists could transfer big packages of data from their experiments. The huge data sets generated in high energy physics experiments also have pushed the envelope in new computing capabilities.

The Tevatron was making a last-ditch effort to find the Higgs boson, the so-called God particle, before its long-planned shutdown in 2011. The Higgs was discovered in July 2012 at CERN, which succeeded Fermilab at the energy frontier with its new Large Hadron Collider (LHC). Many U.S. particle physicists now take part in experiments at the LHC, working from remote control rooms at Fermilab as well as traveling to CERN.

Fermilab has continued to explore other particle realms like neutrinos and dark matter. Thanks to a last-minute deal in Congress, it narrowly averted $36 million in cuts as a result of the drastic across-the-board budget reductions known as "sequestration" in the spring of 2013.

But sequestration hasn't been the only threat to pure research. Since 1970, funding in the physical sciences has been cut in half as a percentage of the gross domestic product. In an April 2013 speech to the NAS, President Barack Obama outlined a proposal to double the funding for the Department of Energy's Office of Science, currently about $5 billion (less than 1 percent of the defense base budget).

"At such a difficult moment," Obama said, "there are those who say we cannot afford to invest in science, that support for research is somehow a luxury at moments defined by necessities. I fundamentally disagree. Science is more essential for our prosperity, our security, our health, our environment, and our quality of life than it has ever been before."

Obama said that as a result of the steady decline in investment since the end of the Moon landing program in the 1970s, "other countries are now beginning to pull ahead in the pursuit of this generation's great discoveries." He declared that "the enormous investment in that era—in science and technology, in education and research funding—produced a great outpouring of curiosity and creativity, the benefits of which have been incalculable . . . We have to replicate that."

"The private sector is unlikely to invest in pure research," Obama said, because the results are too uncertain. That's why "the public sector must invest in this kind of research—because while the risks may be large, so are the rewards for our economy and our society."

That reasoning was crystallized by Fermilab founding director Robert R. Wilson when he testified before the Congressional Joint Committee on Atomic Energy in 1969. He was asked about the value of high energy physics research in the support of national defense. "It has nothing to do directly with defending our country," Wilson said, "except to help make it worth defending."

Mike Perricone is a science writer in Riverside, IL. He has published a book on cosmology, The Big Bang *(Chelsea House, 2009), as well as a collection of personal essays,* From Deadlines to Diapers: Journal of an At-Home Father *(Noble Press, 1992). He has also worked as a sportswriter at the* Chicago Sun-Times, *and as senior editor in Public Affairs at Fermi National Accelerator Laboratory.*

LYSENKO'S REVENGE: SCIENCE, POLITICS, AND HOT RHETORIC
John Galbraith Simmons

More than half a century ago, Trofim Lysenko found himself stripped of his authority over biology and agricultural science in the former Soviet Union. He died about 10 years later, in 1974. When he had arrived in power, through a series of political appointments in the 1930s, he derided genetics in favor of his own theories, destroyed a generation of biologists—many of whom ended up murdered or in prison—and caused untold grief through famine that owed in some measure to his policies. Scientists worldwide tend to prize the story of how Lysenko used ideology to determine and defend false science unsupported by experiment, viewing his rise and fall as an object lesson in how wishful thinking and political power can pervert normal science, infest policy, and lead to calamitous consequences for millions of people (Simmons 1996; Soyfer 2003).

Power is the nexus at which both science and politics intersect and affect the everyday world. Politics is fundamentally about belief systems and "who does what to whom." But with its long reach via technology into human affairs, beginning with the first Industrial Revolution, science has played an increasingly central role in the sphere we call today public policy. Science makes it possible to extensively understand and manipulate nature and also to recognize the limits of that understanding and inflection of power. In that way experiment and quantification can overcome the weight of custom and traditional beliefs and shape economic life. The power science acquires includes economic heft, and with its acquisition comes the prospect that individuals or groups may be able to subvert and corrupt it to selfish or ideological ends. Corrupt power was at the heart of Trofim Lysenko's reign over biology in the Soviet Union beginning in the 1930s; his name abides as a term of derision.

Today, both sides of an unfortunate debate on climate change and global warming invoke Lysenkoism. Its original ex parte use came with respect to biology, relevant beginning in the 1980s, when proponents of newly minted brands of "creation science" attempted to legitimize religious beliefs that contested the evolution of species and compel discussion of those beliefs in high school and college biology curricula. Although "intelligent design" remains a belief system for many, their efforts to intrude on education have largely been failures. In consequence of the constitutional separation of church and state, "creationists" lost repeatedly in the U.S. courts. And the simple fact is that to deny the reality of Darwinian evolution in today's world is like joining the Flat Earth Society; it is entirely without scientific foundation.

With climate change, the situation is somewhat distinct.

The scientific consensus that validates global warming is both real and robust, despite a continual din and outbursts of controversy in the popular media (Oreskes 2004). However, it is fairly recent. Just 20 years ago, an authoritative popular source could assert—although such a claim was tendentious even then—that no such consensus existed (Barnes-Svarney 2004). The mid-1990s in fact saw consolidation of the hypothesis that anthropogenic (man-made) global warming represents a genuine phenomenon that will have serious, perhaps highly destructive consequences in decades to come.

But Earth and climate science today, as multidisciplinary efforts to model global warming, are complicated, complex, multidisciplinary, and evolving. Complexity is probably one reason that some prominent scientists remain skeptics. In 2010, for example, Harold Lewis, a professor emeritus at the University of California, resigned from the American Physical Society, citing that organization's statement on global warming; he objected in particular to the use of the word "incontrovertible" to describe the evidence in its favor (Lewis 2010). The famous physicist Freeman Dyson has also voiced skepticism (Dyson 2008). Nothing about this should be disconcerting. To take an example from another complex field, brain science, not everyone believed that neurons exist, even long after they were definitively demonstrated and typed at the turn of the 20th century. In 1949, nearly 50 years later, the famous neurophysiologist Rita Levi-Montalcini recalled meeting the "reticularist" Han Boeke in Chicago. The skeptic's deep blue eyes, she wrote, "with their somewhat fanatical expression and his proud demeanor, brought to mind the last survivors of Garibaldi's army whom as I child I had seen parading in wheelchairs and leaning on sticks, wearing their characteristic jackets and red berets, in the patriotic processions of 1918" (Levi-Montalcini 1988).

The charge of Lysenkoism with respect to global warming seems to have originated, paradoxically, with climate change skeptics. Sixteen individuals, most but not all scientists, at least one relatively young but a solid majority 70 years old or older, published a letter in the *Wall Street Journal* in 2012 that contested global warming. They complained that they could not get a fair hearing and recalled when "Trofim Lysenko hijacked biology in the Soviet Union" ("No Need to Panic" 2012). Spinning their convictions into the political sphere, their letter was directed to "any candidate for public office." However, because Trofim Lysenko rejected the scientific consensus (fairly recent in the 1930s) concerning genetics, the climate skeptics' charge had a poor fit with their own views and with history itself. Not surprisingly, soon they themselves were being labeled Lysenkoists (Mooney 2012).

After its appearance in the *Wall Street Journal*, the term Lysenkoism reverberated through the blogosphere. Here the problem of ignorance takes a toll on the credibility of skeptics owing to their expansive and simplistic rhetoric; but here also it potentially reaches an intelligent but vulnerable audience in the business community—and others, including some journalists (Freudenburg and Muselli 2010). To take one example, Peter Ferrara, a lawyer and former official with the Reagan and Bush administrations, writes in a blog for *Forbes* magazine that "[f]ormerly respected scientific bodies in the U.S. and other western countries have been commandeered by political activist Lysenkoists seizing leadership positions" (Ferrara 2013). He describes *Science* and *Nature*, the premier science journals in the United States and Britain, respectively, as "once respected" and as "having abandoned science for Lysenkoism." For reliable information he directs his readers to a report by the Cato Institute, a libertarian think tank. The charge of anthropogenic global warming (AGW) in American science, he writes, is "an infestation."

At root the overriding problem for climate-change skeptics—the reason their writings are largely confined to the blogosphere, the "gray literature" of think tanks, Internet newsletters, and industry journals—is that the underlying science is settled. The basic chemistry behind carbon dioxide–induced warming actually dates to the 19th century. French scientists discovered the ozone layer in the early 20th century and the role of synthetic chemicals in its depletion dates to the 1970s. Although the factors that determine the anthropogenic contribution to global warming are complex, the basic science is well understood. "Nobody can publish an article in a scientific journal claiming the Sun orbits the Earth," writes Naomi Oreskes, "and . . . you can't publish an article in a peer-reviewed journal claiming there's no global warming" (Oreskes and Conway 2010).

Hence the skeptic's recourse to Lysenko. Peter Ferrara's language, the high dudgeon of this rhetoric, tells the story best. "My how western science has fallen," he writes. He thinks that "the day the theory of anthropogenic catastrophic global warming died" was the day the issue was debated at the Heartland Institute with which he is associated. He provides a strikingly inept characterization of science itself: "Real science . . . is not a matter of consensus," he writes, "but of reason, with skepticism at its core." That would be a good description of the philosophy of David Hume, but not the science of Isaac Newton, much less Albert Einstein. Quite like Lysenko himself, he derides normal science and alludes to but does not specify "better recent science" by which "the naturally stable Earth would enjoy negative feedbacks restoring long term equilibrium and stability to global temperatures" (Ferrara 2013).

In the end, the newly fashioned charge of Lysenkoism, promulgated first and in large measure by a small group of mostly elderly scientists and global warming skeptics, is unsustainable.

References

Barnes-Svarney, P.L. 2004. *The New York Public Library Science Desk Reference*. New York: Macmillan, xix.

Dyson, F. 2008. "The Question of Global Warming." *New York Review of Books*. 55(14): 92.

Ferrara, P. "The Disgraceful Episode of Lysenkoism Brings Us Global Warming Theory." http://www.forbes.com/sites/peterferrara/2013/04/28/the-disgraceful-episode-of-lysenkoism-brings-us-global-warming-theory/. Accessed on September 25, 2013.

Freudenburg, W.R., and V. Muselli. 2010. "Global Warming Estimates, Media Expectations, and the Asymmetry of Scientific Challenge." *Global Environmental Change*. 20(3): 483–491.

Levi-Montalcini, R. 1988. *In Praise of Imperfection: My Life and Work*. New York: Basic Books, xiii.

Lewis, H. 2010. "Hal Lewis: My Resignation from the American Physical Society." http://www.thegwpf.org/hal-lewis-my-resignation-from-the-american-physical-society/. Accessed on September 25, 2013.

Mooney, C. 2012. "In Which Climate 'Skeptics' Drop the Lysenko Bomb. No, I'm Not Kidding. . . ." http://www.desmogblog.com/which-climate-skeptics-drop-lysenko-bomb-no-i-m-not-kidding. Accessed on September 2, 2013.

"No Need to Panic about Global Warming." *Wall Street Journal*. January 23, 2012.

Oreskes, N. 2004. "The Scientific Consensus on Climate Change." *Science*. 306(5702): 1686.

Oreskes, N., and E.M. Conway. 2010. *Merchants of Doubt: How a Handful of Scientists Obscured the Truth on Issues from Tobacco Smoke to Global Warming*. New York: Bloomsbury Press.

Simmons, J.G. 1996. *The Scientific 100: A Ranking of the Most Influential Scientists, Past and Present*. Secaucus, NJ: Carol Publishing Group, xix.

Soyfer, V.N. 2003. "Tragic History of the VII International Congress of Genetics." *Genetics*. 165(1): 1–9.

John Galbraith Simmons writes about science and medicine for both general and professional audiences. He is the author of The Scientific 100 *(Citadel, 1996),* Doctors and Discoveries: Lives That Created Today's Medicine *(Houghton Mifflin, 2002), and most recently, in collaboration with Justin Zivin, MD, Phd,* tPA for Stroke: The Story of a Controversial Drug *(Oxford, 2010).*

THE CORRUPTION OF CLIMATE SCIENCE BY LEFTIST POLITICS
A. J. Smuskiewicz

"Political influences in government-sponsored research have focused climate change research on CO_2 rather than a broader range of factors that need better definition," wrote a group of former NASA astronauts, scientists, and engineers in an April 2013 report titled "Anthropogenic Global Warming Science Assessment Report" (The Right Climate Stuff Research Team; Berger).

That unusually blunt statement helps us begin to understand how climate scientists have become corrupted and co-opted by the empowered political left. The government bureaucrats and administrators who bestow research grants have the power to decide what kind of science gets funded and what kind is left penniless. As the report suggests, scientists who want to look into natural causes or less catastrophic outcomes for climate change—rather than apocalyptic, anthropogenic (human-caused) global warming—often see their proposals go unfunded (The Right Climate Stuff Research Team; Berger).

Is it possible that the federal agencies in charge of funding environmental and climate science have a preconceived political agenda that tends to lean toward the left (i.e., in favor of more government regulation)? Well, there is certainly nothing like a human-caused environmental disaster to build the case for more government power.

Money is another corrupting influence in climate science. The "green" industry has become very profitable. First among the environmentalist millionaires is former vice president Al Gore. His venture capital investments in green-energy technologies that have also received government support have yielded huge profits (Broder; Global Warming Winners). It has been said that Gore stands to become the world's first "carbon billionaire" as a result of his investments in the multibillion-dollar carbon credit trading market, which was designed to use market forces to reduce carbon dioxide emissions. But Gore's hypocrisy regarding the environment was evident in January 2013 when he sold his struggling cable television station, Current TV, to Al Jazeera, the Islamic network funded by the oil-rich government of Qatar, for $500 million (Elias; Glover).

Many climate scientists are also involved with lucrative investments in alternative energy ventures that they profit from as a result of government policies for which they advocate (Global Warming Winners). It is not only "Big Oil" that has a strong financial interest in the climate change debate.

The corruption resulting from the combination of climate science and leftist politics is most evident in the United Nations' Intergovernmental Panel on Climate Change (IPCC), an organization commonly hyped by the media as the most authoritative scientific body on climate change (Worldgate). The IPCC regularly issues dire reports on climate change that are highly influential among government policy makers. However, environmental science researcher Denis Rancourt characterizes the IPCC as a group of scientists who "accept to serve a political role" (Morano a). Rancourt, who happens to be a leftist, calls the anthropogenic global warming movement a "corrupt social phenomenon" and "comfortable lies" designed to "alleviate the guilt that they [Western environmentalists] feel about being on the privileged end of the planet."

The biases of the IPCC became embarrassingly clear in 2009 when hundreds of hacked e-mails by scientists associated with the UN panel were made public (Terrell; Worldgate). The e-mails revealed that scores of scientists purposely conspired to fraudulently report climate data. They spoke of the need to "hide" data that did not support anthropogenic global warming—specifically by not publishing those studies in the IPCC reports "even if we have to redefine what the peer-review literature is!"

One of the IPCC scientists participating in this e-mail conspiracy was Michael Mann, the meteorologist who developed the famous "hockey stick" graph that purports to show stable global temperatures between AD 1000 and 1900, when temperatures allegedly began a dramatic rise caused by industrial pollution (Daly; Terrell; Worldgate). This graph, which Mann based on tree ring analyses, continues to be widely cited in school texts and by politicians, including Gore, as proof of anthropogenic global warming. The problem is that this graph is riddled with scientific flaws, including the deletion of a previously established Medieval Warm Period and the ignoring of satellite data on temperatures, which are at odds with the tree ring data (Daly; Terrell; Worldgate).

Mann's shaky hockey stick points to some of the problems with computer climate models, which form the foundation of many climate change studies. When working with computer models, if you put garbage data in, you get garbage results out. Or as renowned physicist Freeman Dyson wrote in his autobiography, ". . . all the fuss about global warming is grossly exaggerated. . . . I have studied the climate models and I know what they can do. . . . They do a very poor job of describing the clouds, the dust, the chemistry and the biology of fields and farms and forests. They do not begin to describe the real world that we live in" (Lewis).

The problems with climate models were highlighted by the previously mentioned group of former NASA astronauts, scientists, and engineers—a group called The Right Climate Stuff (TRCS) Research Team (Berger; The Right Climate Stuff Research Team). In their report, these researchers noted that the actual response of Earth's surface temperatures to carbon dioxide emissions has never been validated in models. Such validation would require decades of real-world observation and data collection, followed by the establishment of formal protocols on which to build the models.

The TRCS report reached the following conclusions:

- "Carbon-based AGW [anthropogenic global warming] science is not settled."

- "Natural processes dominate climate change (although many are poorly understood)."

- "Empirical evidence for Carbon-based AGW does not support catastrophe."

The TRCS group is not alone in its science-based skepticism of anthropogenic global warming Armageddon. A 2012 poll of professional meteorologists belonging to the American Meteorological Society (AMS) found that only 30 percent were "very worried" about global warming (Taylor a; Taylor b). And a 2010 poll of AMS meteorologists who worked as weather forecasters on television found that only 24 percent believed that global warming was caused by human activities, and only 19 percent believed that climate models were reliable (Taylor a; Taylor c). These findings of AMS member opinions stand in stark contrast to the official position of the AMS leadership, which is firmly in the doomsday camp (Taylor b; Taylor c). This is one of many indications that the leadership of the scientific establishment is concerned mainly with politics, while the rank-and-file members do the real science.

When skeptical scientists speak out, they risk professional retaliation by this establishment. For example, when meteorologist Henk Tennekes questioned the scientific basis of global warming, he was fired as research director of the Royal Dutch Meteorological Society (Lewis). When botanist and television personality David Bellamy announced his skepticism, he found his career at the BBC coming to a sudden end (Morano a). Denis Rancourt says that his skeptical comments have caused publishers to shun his articles (Morano a). This is a form of scientific censorship and enforced silence that makes it hard for the public to hear these skeptical views.

The apparent ignorance of the newsreaders and commentators in the mass media does not contribute to an intelligent public discussion regarding climate change. Many of these "journalists" seem to lack the historical or scientific knowledge to put the latest hurricane or heat wave in proper perspective. Are they aware that the 1930s were one of the warmest decades on record (Climate Skeptic; Moran b)? Or that the 1950s were a time of unusually intense hurricane activity (Climate Skeptic; Hurricanes)? Or that the "scientific consensus" during the 1970s was that Earth was entering a new Ice Age (Climate Skeptic; Sutton)? Or that surface temperature and ocean heat records from two large, independent databases suggest that there has been little or no surface warming since 1995 (Is the Heat Hiding?; McGrath; Meyer; Where's the Acceleration?)? Do they even know that climate, by its very definition, is always changing?

Ultimately, we as individuals who are interested in the truth need to exercise our own skills of critical thinking when we absorb the mass of information and disinformation that gets thrown at us from every direction. Then we need to do our own investigation.

References

Berger, Eric. "Apollo-era NASA Officials Say Climate Change Research 'Corrupted' by Politics and Special Interests." Houston Chronicle blogs. http://blog.chron.com/sciguy/2013/04/apollo-era-nasa-officials-say-climate-change-research-corrupted-by-politics-and-special-interests/. Accessed on August 26, 2013.

Broder, John M. "Gore's Dual Role: Advocate and Investor." *New York Times.* http://www.nytimes.com/2009/11/03/business/energy-environment/03gore.html?_r=0. Accessed on August 26, 2013.

"Climate Skeptic." http://www.climate-skeptic.com/. Accessed on August 26, 2013.

Daly, John L. "The 'Hockey Stick': A New Low in Climate Science." http://www.john-daly.com/hockey/hockey.htm. Accessed on August 26, 2013.

Elias, Paul. "Al Gore Sued over Current TV Sale to Al Jazeera." *Huffington Post*. http://www.huffingtonpost.com/2013/03/07/al-gore-sued-current-sale-al-jazeera_n_2827366.html. Accessed on August 26, 2013.

"Global Warming Winners." *The Washington Times*. http://www.washingtontimes.com/news/2010/mar/3/global-warmings-biggest-winners/#0_undefined,0_. Accessed on August 26, 2013.

Glover, Peter C. "Green Hypocrisy as Al Gore Sells out for Petro-dollars." *Energy Tribune*. http://www.energytribune.com/70283/green-hypocrisy-as-al-gore-sells-out-for-petro-dollars#13757183564181&95672::resize_frame|124–140. Accessed on August 26, 2013.

"Hurricanes: The Greatest Storms on Earth." *NASA Earth Observatory*. http://earthobservatory.nasa.gov/Features/Hurricanes/hurricanes_3.php. Accessed on August 26, 2013.

"Is the Heat Hiding? Ocean Heat Content Hasn't Risen." Climate Skeptic. http://www.climate-skeptic.com/wp-content/uploads/2012/04/ocean-heat.gif. Accessed on August 26, 2013.

Lewis, James. "'Global Warming' as Pathological Science." *American Thinker*. http://www.americanthinker.com/2007/11/global_warming_as_pathological.html. Accessed on August 26, 2013.

McGrath, Dan. "Global Warming? Temperature up 'Very Close to Zero' over 15 Years." GlobalClimateScam.com. http://www.globalclimatescam.com/category/ipcc/. Accessed on August 26, 2013.

Meyer, Warren. "A Vivid Reminder of How the Climate Debate is Broken." *Forbes*. http://www.forbes.com/sites/

warrenmeyer/2012/04/19/a-vivid-reminder-of-how-the-climate-debate-is-broken/. Accessed on August 26, 2013.

Morano, Marc (a). "Left-wing Env. Scientist Bails out of Global Warming Movement: Declares It a 'Corrupt Social Phenomenon . . . Strictly an Imaginary Problem of the 1st World Middleclass.'" Climate Depot. http://www.climatedepot .com/2010/07/26/leftwing-env-scientist-bails-out-of-global-warming-movement-declares-it-a-corrupt-social-phenomenon strictly-an-imaginary-problem-of-the-1st-world-middle class/. Accessed on August 26, 2013.

Morano, Marc (b). "Update: 'Hottest Year' in U.S. Called into Question—1930s Still Reign as Hottest?—Global Temps Remain Stagnant." Climate Depot. http://www.climatedepot .com/2013/01/09/update-hottest-year-in-us-called-into-question-1930s-still-reign-as-hottest-global-temps-remain-stagnant/. Accessed on August 26, 2013.

The Right Climate Stuff Research Team. Anthropogenic Global Warming Science Assessment Report. http://www.therightcli matestuff.com/AGW%20Science%20Assess%20Rpt-1.pdf. Accessed on August 26, 2013.

Sutton, Gary. "The Fiction of Climate Science." *Forbes.* http:// www.forbes.com/2009/12/03/climate-science-gore-intelligent-technology-sutton.html. Accessed on August 26, 2013.

Taylor, James (a). "Meteorologists Reject U.N.'s Global Warming Claims." http://news.heartland.org/newspaper-article/2010/02/01/meteorologists-reject-uns-global-warming-claims. Accessed on August 26, 2013.

Taylor, James (b). "Peer-reviewed Survey Finds Majority of Scientists Skeptical of Global Warming Crisis." *Forbes.* http://www .forbes.com/sites/jamestaylor/2013/02/13/peer-reviewed-sur vey-finds-majority-of-scientists-skeptical-of-global-warming-crisis/. Accessed on August 26, 2013.

Taylor, James (c). "Shock Poll: Meteorologists Are Global Warming Skeptics." *Forbes.* http://www.forbes.com/sites/

jamestaylor/2012/03/14/shock-poll-meteorologists-are-global-warming-skeptics/. Accessed on August 26, 2013.

Terrell, Rebecca, and Ed Hiserodt. "IPCC Researchers Admit Global Warming Fraud." *The New American.* http://www
.thenewamerican.com/tech/environment/item/6748-ipcc-researchers-admit-global-warming-fraud. Accessed on August 26, 2013.

"Where's the Acceleration? Temperatures Have Been Flat for a Decade." Climate Skeptic. http://www.climate-skeptic.com/wp-content/uploads/2012/04/recent-temp.gif. Accessed on August 26, 2013.

"Worldgate, the Incredible History of the IPCC: The Science of Lying Now Hegelian Tactics." Euro-Med. http://euro-med
.dk/?p=13721. Accessed on August 26, 2013.

A. J. Smuskiewicz is a biologist and a freelance writer specializing in science and medicine. He has more than 20 years' experience producing a variety of educational articles and books, including a book about climate change for school children.

OIL AND WATER—THE POLITICAL MIX OF THE 21ST CENTURY
Lana Straub

Hydraulic fracturing is becoming one of the most contro-versial technologies of the 21st century. However, it is not a new science. According to the article "Hydraulic Fracturing, A History of Enduring Technology," "[t]he first experimental treatment to 'hydrofrac' a well for stimulation was performed in the Hugoton gas field in Grant County, Kansas in 1947 by Stanolind Oil" (Montgomery and Smith 2010). The hydraulic fracturing process was used in the oil exploration business for over six decades before it became a national issue of concern. However, with the invention of instantaneous media sharing, the inner workings of the process and its potential for causing contamination became known to the common citizen through various media outlets, including movies. Public knowledge of widespread use of the science of hydraulic fracturing has caused great concern and grassroots uprising and has divided the United States across political, moral, and ethical boundaries.

It is common knowledge that petroleum products come from mining fractures found in Earth's surface. The science of hydraulic fracturing "involves the injection of fluids under pressures great enough to fracture the oil-and-gas producing formations. The resulting fractures are held open using 'proppants,' such as fine grains of sand or ceramic beads, to allow oil and gas to flow from small pores within the rock to the production well" ("EPA's Study of Hydraulic Fracturing and Its Potential Impact on Drinking Water Resources" 2013). This technology has allowed companies to return to previous sites once rich in natural resources gone dormant, and rejuvenate those sites with chemical additives allowing oil and gas to flow once again. As a science, it has revolutionized the oil and gas drilling industry, creating a new oil and gas boom in the new millennium. It is the hope of many lawmakers, business owners, and citizens that the hydraulic fracturing science will lead to American energy independence.

The other side of energy independence is the use of chemicals in the process of hydraulic fracturing. The question as to the effects of hydraulic fracturing on potential drinking water sources, whether they are groundwater or surface water, has become a hot button topic for many in the United States. Proponents of hydraulic fracturing warn that the public is uninformed or misinformed about the truth behind the science. Opponents maintain that more studies need to be done before fracking can be considered safe. These polar opposite ideas have caused a great deal of controversy in the United States. On one side of the argument are the proponents of energy independence and economic rejuvenation, who say that hydraulic fracturing brings economic recovery to America. On the other side are those who argue that the environmental impact of chemicals on the land and clean drinking water sources should be paramount.

The controversy and conflict of economic gain and rejuvenation versus public safety and environmental stewardship have caused parts of the country to become polarized by this politically charged issue. As all politics is local and grassroots movements begin with the effect of a problem on a locality, so goes the movement to halt hydraulic fracturing and to urge the Environmental Protection Agency to study the effects of hydraulic fracturing on drinking water supplies. The loudest outcry has come from the almost 9 million citizens who take their water from the Delaware River Basin.

Public outcry caused New York State to put a moratorium on hydraulic fracturing and elicit the help of the U.S. Congress to study all of the possible effects of the processes involved in hydraulic fracturing and the effects those processes might have on drinking water. The U.S. House of Representatives Committee on Energy and Commerce is still conducting studies of the science. The Environmental Protection Agency (EPA) has also undertaken a study. After beginning its study, the EPA realized the daunting task associated with monitoring all of the different procedures involved in the hydraulic fracturing process. Realizing the breadth of such an undertaking, the agency decided to focus a study on any possible relationships between hydraulic fracturing and drinking water contamination. The EPA Plan to Study the Potential Impacts of Hydraulic Fracturing ("EPA's Study of Hydraulic Fracturing and Its Potential Impact on Drinking Water Resources" 2013) began in 2011 and was slated for completion in 2013, but has been delayed. The EPA hopes it will be completed in 2014, barring any more delays.

Several movies have been made addressing/fueling the public's concern over fracking. *Gasland the Movie* (Directed by Josh Fox 2010) and *Frack Nation* (Directed by Phelim McAleer, produced by Hard Boiled Films 2013) are two big documentaries that discussed the issue. *Promised Land* (Directed by Gus

Van Sant, produced by Matt Damon and Francis McDormand 2012) was a fictional tale starring Matt Damon aimed at exposing deception of oil companies. "The *New York Times* has published many articles on the fracking issue," states Sarah Meyland, director of the Center for Water Resources Management at the New York Institute of Technology. "Also, the problems that have been experienced in Pennsylvania and elsewhere have become well-known here," She continues. "The movie, *Gasland*, has been widely viewed and has helped shape public opinion" (Meyland 2013).

Like it or not, the world's advanced nations are heavily dependent on fossil fuels to power their business economies and their personal lives. The petroleum industry hopes that hydraulic fracturing can survive the political turmoil as they feel the technology is the key to energy independence. "As the global balance of supply and demand forces the hydrocarbon industry toward more unconventional resources including U.S. shales such as the Barnett, Haynesville, Bossier and Marcellus gas plays, hydraulic fracturing will continue to play a substantive role in unlocking otherwise unobtainable reserves" (Montgomery and Smith 2010). States like New York disagree. While the EPA's study is ongoing at the time of this writing, so is New York's independent environmental study. The New York State Assembly has voted to extend its moratorium on hydraulic fracturing until at least May 15, 2015 while it awaits the findings of a New York Department of Health study on the issue.

While other states follow New York's lead and some others choose not to do so, the United States citizenry remains on this issue, as on many other issues in the 21st century, a nation divided between the search for economic and energy independence and public health and environmental safety. "We need to develop our resources as continued independence and security," states Barry Stephens, president of TBD America Inc., a global technology business development consulting group serving the public and private sectors in the energy, fuels, and water treatment industries. "We need U.S. oil," maintains Stephens, "there is no doubt about it" (Stephens 2013).

While some people maintain that hydraulic fracturing is completely safe, some Americans still believe that oil and water don't mix. "People fear the consequences of hydraulic fracturing will exceed the benefits and the economics is tied up with the science," states Meyland. "A lot of people see articles where companies leave a site and the site is spoiled for other uses," she states, "it makes people doubly fearful and careful" (Meyland 2013).

The political debate is likely to rage on for many more years as the country remains polarized on this issue, at least until the results of the studies come out to confirm or deny the effects of science on the environment.

References

"EPA's Study of Hydraulic Fracturing and Its Potential Impact on Drinking Water Resources." U.S. Environmental Protection Agency. http://www2.epa.gov/hfstudy. Accessed on September 29, 2013.

Meyland, Sarah. Director, Center for Water Resources Management. New York Institute of Technology. Interview by Lana Straub. September 19, 2013.

Montgomery, Carl T., and Michael B. Smith. 2010. "Hydraulic Fracturing: History of an Enduring Technology." *Journal of Petroleum Technology*. 62(12): 26–32.

Stephens, Barry. President, TBD America, Inc. Interview by Lana Straub. September 6, 2013.

Lana Straub is a graduate of Texas A&M University—Commerce, with a degree in political science. She has been a freelance journalist since 2002.

4 Profiles

Introduction

One way of understanding the interaction of science and politics is by learning more about the individuals and organizations that are and have been involved in that process. Learning more about the thought and work of historic figures such as Plato, Sir Francis Galton, Philipp Lenard, and B. F. Skinner, as well as modern figures such as Jonathan Beckwith, Edward Korman, John Marburger, and Eugenie Scott, provides an insight into the ways that scientists think about political issues, and how politicians think about and use science. Studying the work of organizations such as the American Association for the Advancement of Science, the Coalition for the Public Understanding of Science, the Federation of American Scientists, and the Government Accountability Project reveals the way in which groups of interested individuals can plan together to prevent the misuse of science and/or politics. This chapter provides sketches of such individuals and organizations. Brief biographical notes are also provided on other individuals who are perhaps themselves not so widely known, but whose thought, philosophy, and actions allow the student of science and politics issues to understand the motivations that individuals may have and the actions they

Health and Human Services Secretary Kathleen Sebelius speaks about the Health Insurance Marketplace at the Community Health and Social Services Center in Detroit in November 2013. (AP Photo/Paul Sancya)

take with regard to this field of human endeavor. In such cases, the individuals themselves may or may not be major figures in this story, but their public representations about important scientific and political issues probably are of considerable interest.

Todd Akin (1947–)

> Concern for rape victims is a red herring because conceptions from rape occur with approximately the same frequency as snowfall in Miami.
>
> James Leon Holmes, Chief Judge, Eastern District of Arkansas

> The facts show that people who are raped—who are truly raped—the juices don't flow, the body functions don't work and they don't get pregnant. Medical authorities agree that this is a rarity, if ever.
>
> Henry Aldridge (R), North Carolina House of Representatives

> The odds of a woman becoming pregnant through rape are "one in millions and millions and millions." The trauma of rape causes women "to secrete a certain secretion" which has a tendency to kill sperm.
>
> Stephen Freind (R), Pennsylvania House of Representatives

For at least two decades, a theory has been popular among some politicians in state legislatures and the U.S. Congress about the risk of a woman's becoming pregnant as the result of rape. As these comments indicate, that theory suggests that a woman's body responds in some way to the trauma of rape and produces a secretion not normally present in the body that eliminates the possibility of pregnancy's occurring. This theory has been subjected to scientific scrutiny on a number of occasions, and there appears to be no evidence in its support. Still, politicians continue to offer the theory for a variety of reasons.

During the 2012 national elections, the theory appeared once more, this time during the race of U.S. Senate in the state of Missouri between incumbent Claire McCaskill (MO-D) and challenger Todd Akin (R). When queried about his strong opposition to abortion, even in cases of rape, Akin explained "From what I understand from doctors, that's [a woman's becoming pregnant] really rare. If it's a legitimate rape, the female body has ways to try to shut that whole thing down." Akin was challenged by critics not only for invoking, once again, the outdated theory about pregnancy and rape, but also for suggesting that there might be both "legitimate" and "illegitimate" forms of rape. In any case, Akin's statement appeared to have had a profound effect on the race. In the November election, he was defeated by Senator McCaskill, 54.7 percent to 39.2 percent, in a contest that Akin had been expected to win with ease prior to his statement.

William Todd Akin was born in New York City on July 5, 1947, the son of Paul Bigelow Akin and Nancy Perry Bigelow Akin. The Akin family moved to St. Louis, where Todd attended the John Burroughs School, a private academy, before matriculating at Worcester Polytechnic Institute, in Worcester, Massachusetts. He earned his bachelor's degree in engineering from Worcester in 1971, before joining the U.S. National Guard, where he served as an engineer. He then continued to serve in the Army Reserve until 1980.

After leaving the army, he enrolled at the Covenant Theological Seminary in St. Louis, from which he earned a master of divinity degree in 1984. Although he never joined the ministry, he did become very active in causes of special interest to him, such as working in opposition to abortion rights for women. He was eventually arrested eight times for protesting at abortion clinics in the Midwest.

Akin's first entry into politics occurred in 1988 when he ran unopposed for a seat in the Missouri House of Representatives. He won reelection to the seat five more times before standing for election to the U.S. House of Representatives from Missouri's second district. He won that election in 2000 with 55 percent of the vote, and was reelected five more times, with majorities ranging from 61 percent to 68 percent of the popular vote.

Akin's defeat in the 2012 senatorial race was particularly disappointing for the National Republican Party because of its earlier hopes and expectations of taking control of the Senate from the Democratic Party. Ironically, the Republicans lost a second seemingly "sure bet" Senatorial election in the State of Indiana, where Republican candidate Richard Mourdock apparently lost an early lead after commenting on the topic of rape. As horrible as rape is, he had said, he had come to understand that "life is that gift from God that I think even if life begins in that horrible situation of rape, that it is something that God intended to happen." Enough voters disagreed with Mourdock on this point that he lost the election to U.S. Representative Joe Donnelly (D) by a vote of 50.0 percent to 44.3 percent. It would appear, then, that the Republic Party may have lost an opportunity to take control of the U.S. Senate in 2012 because of two controversial comments about the biological facts of rape.

American Association for the Advancement of Science

1200 New York Ave., NW
Washington, DC 20005
Phone: (202) 326–6400
E-mail: http://www.aaas.org/contact.shtml
URL: http://www.aaas.org/

The American Association for the Advancement of Science (AAAS) is one of the oldest, largest, and most respected all-purpose scientific societies in the world. It was founded on September 20, 1848, as an outgrowth of the Association of American Geologists and Naturalists for the purpose of "promot[ing] intercourse between those who are cultivating science in different parts of the United States, to give a stronger and more general impulse, and a more systematic direction to scientific research in our country; and to procure for the labours of scientific men, increased facilities and a wider usefulness." Among the association's 87 founding members were the elite of the U.S. scientific community at the time, including Alexander Dallas

Bache, great-grandson of Benjamin Franklin and head of the U.S. Coast Survey, Louis Agassiz, Joseph Henry, Benjamin Peirce, Henry Darwin Rogers and his brother William Barton Rogers, James Dwight Dana, Oliver Wolcott Gibbs, Benjamin A. Gould, William Redfield, and Benjamin Silliman, Jr.

Today AAAS is affiliated in one way or another with 261 specialized societies and academies with more than 10 million members worldwide. Its primary journal, *Science*, has the largest paid subscription of any peer-reviewed scientific journal in the world. It was founded in 1880 and has been publishing without interruption ever since. The journal includes reports and reviews of research, science-related news, opinion pieces on selected issues related to science, letters to the editor, and reports of scientific meetings and conferences. It differs from the vast majority of other scientific organizations in that it remains, as its founders had intended, a general interest society with information about and by scientists in every conceivable field.

Much of the work of the association is carried out through its 24 sections, including agriculture, food, and renewable resources; anthropology; astronomy; atmospheric and hydrospheric sciences; biological sciences; chemistry; dentistry and oral health sciences; education; engineering; general interest in science and engineering; geology and geography; history and philosophy of science; industrial science and technology; information, computing, and communication; linguistics and language science, mathematics, medical sciences, neuroscience, pharmaceutical sciences; physics; psychology; social, economic, and political sciences; societal impacts of science and engineering; and statistics.

An important part of AAAS' work reflects one of its initial goals: educating government about issues that have a science component. In its document, *AAAS Policy, Guidelines, and Procedures for Communications with Congress*, the association makes clear that it does not engage in activities that would be regarded as political in nature. It does not provide financial support, equipment, supplies, or communications or physical facilities for any type of political campaign on any level, federal, state, or local. AAAS makes very clear that it takes part in no direct or grassroots lobbying of any kind. The thrust of this policy is that the organization does not take positions on controversial issues, such as climate change, abortion, or food policy.

By contrast, the association sees as its role providing information to the U.S. Congress, the Executive Branch, and other governmental entities, when asked, about the scientific content of important issues. With the very large and experienced collection of scientists at its disposal, AAAS is probably better qualified than almost any other nongovernmental agency that one could imagine for analyzing important national problems and offering a variety of options, with possible outcomes from each option, to decision makers.

AAAS has developed a detailed and comprehensive plan for ensuring that the general guidelines outlined here are actually carried out in practice. Staff members who have occasion to interact with governmental and nongovernmental personnel are required to obtain preapproval for their participation in such activities, and to keep their superiors informed at all times of the nature of their activities with such individuals. These regulations apply to all possible forms of interaction, including oral and written testimony before Congressional committees or task forces; correspondence with Congressional offices of a nonroutine nature; contacts that border on lobbying; and proposals for conducting Congressional seminars. The organization also includes an annual plan of activities that involve the interaction between AAAS staff and Congressional or other governmental staff to ensure that such activities fall within the aforementioned general guidelines.

A number of offices within AAAS are involved in issues related to the interaction of science and government and politics. The Office of Government Relations was established in 1994, for example, for the purpose of providing objective information to the Congress on science-related issues with which it had to deal. The Office publishes a regular newsletter, *Science and*

Technology in Congress, which includes articles of interest to the Congress, as well as regular features such as Congress In Brief, Agency Updates, Reports and Publications, AAAS News and Notes, and Frontiers in Science.

Another program is the R&D Budget and Policy Program, created in 1976. The program sponsors research and public meetings on funding and policy issues affecting support of research and development activities by the federal government. It produces an annual report, *AAAS Report: Research and Development*, which reviews the president's budget proposals for research and development in the coming fiscal year for each of the executive departments and other programs, making use of the special expertise available from specialized science, engineering, and educational societies. The Center for Science, Technology, and Security Policy focuses especially on policy issues dealing with national security, and calls on experts available in and through AAAS to suggest policy developments that enhance the nation's security and examine possible security implications of research being conducted by American scientists.

Jonathan Beckwith (1935–)

At the end of the 1960s, a number of members of the scientific community had become disillusioned with the ways that scientific knowledge was being put to use by politicians in the United States and other nations. This growing concern was related to and part of the general sense of unease that many citizens around the world were feeling about the direction of political movements in many developed nations, as illustrated most clearly by the war being conducted in Vietnam by the United States and its allies against a nationalistic uprising of the Viet Cong movement. As an example, research was being conducted on ways in which weather conditions could be controlled by humans in order to increase the number and intensity of storms in Vietnam, as a way of interfering with Viet Cong movements throughout North and South Vietnam. Concern also focused on the use of sophisticated military weapons and technologies that allowed pilots flying at altitudes of more than 10,000 meters (30,000 feet) to drop bombs and chemical weapons on humans on Earth's surface, many of whom were noncombatants.

Among the many organizations created to express these concerns was a group known as Science for the People (SftP), whose first public statements were issued in 1969. In announcing an organizational meeting of like-minded scientists and nonscientists, organizers explained their position:

> While we now see that many of the products of science and technology have become more a menace than a boon to the interests of human society, the dominant professional associations—such as the APS [American Physical Society]—have deliberately remained aloof from the desperate problems facing mankind today.

In its first newsletter of February 23, 1969, the group outlined its goal as being a way to "provide a means whereby socially aware scientists can act as a coherent body to help solve the vital problems pressing upon us all."

For about a decade, SftP exerted an influence on the thinking of professional scientists, educators, and the general public well out of proportion compared to their actual membership numbers. Their newsletter and later magazine, *Science for the People*, carried some of the most trenchant observations and analyses of the way in which science was being used and some of the strongest calls for changes in the way scientific research should be conducted, supported, and applied in everyday life. One of the earliest members of the group was Jonathan Beckwith, then professor of bacteriology and immunology and professor of microbiology and molecular genetics at the Harvard University Medical School. In the year that SftP was being formed, Beckwith led a research team that made an important breakthrough in the field of molecular genetics, isolating the first gene from a bacterial chromosome.

Jonathan Roger Beckwith was born in Cambridge, Massachusetts, on December 25, 1935. He attended Newton High School, in Newton, Massachusetts, from which he graduated in 1953. He then matriculated at Harvard College, where he was granted his AB in chemistry in 1957 and his PhD in biochemistry in 1961. He served for one year as a teaching assistant at the University of Illinois before completing his postdoctoral studies at the University of California at Berkeley, Princeton University, Hammersmith Hospital in London, and the Institut Pasteur in Paris. In 1965 he accepted an appointment as an associate in the Department of Bacteriology and Immunology at the Harvard Medical School, where he has remained ever since. He continues to direct the work of his laboratory on the *Escherichia coli* bacterium. He is currently American Cancer Society Professor of Microbiology and Molecular Genetics at Harvard Medical School.

Beckwith's list of honors and awards is quite long. It includes the 1970 Eli Lilly Award for Outstanding Achievements in Microbiology, a 1970 Guggenheim Fellowship, appointment as an American Cancer Society Research Professor (1980), a 1986 Merit Award from the National Institutes of Health, 1993 Genetics Society of America Medal, the 2009 Selman Waksman Award in Microbiology of the National Academy of Sciences (NAS), and the 2009 Edinburgh Medal of the city of Edinburgh, Scotland. Beckwith was elected to the NAS in 1984, and was named a fellow of the AAAS in 1985, a fellow of the American Academy of Arts and Science in 1986, and a fellow of the American Academy of Microbiology in 1992. He has served as visiting professor at the University of California at Berkeley, Samuel Rudin Visiting Professor at Columbia University College of Physicians & Surgeons, Lester O. Krampitz Lecturer in Microbiology at Case Western Reserve University, and Roger Stanier Memorial Lecturer at the University of California at Berkeley.

Beckwith has also committed large amounts of time and effort to the pursuit of socioscientific issues. He has, for example, served on the advisory boards of the Council for Responsible Genetics and the Eritrean Relief Committee for nearly 30 years. From 1989 to 1995, he served as a member of the Working Group on Ethical, Legal and Social Implications of the Human Genome Project of the National Institutes of Health. From 1990 to 1993, he served as president of the board of directors of SftP, and since 2002, he has been a member of the advisory board of the Program in Science, Technology and Society Program at the Kennedy School of Government at Harvard University.

Beckwith has published more than 300 peer-reviewed scientific papers on gene expression, the mechanism of protein secretion, the structure and function of membrane proteins, disulfide bond formation in proteins, and cell division, as well as more than 80 papers on topics dealing with the interaction of science and society. Among the most recent of the latter group of papers are "Deconstructing Violence (*GeneWatch,* 2007)," "Twin Studies of Political Behavior: Untenable Assumptions?" (*Perspectives on Politics,* 2008), and "Illusions of Scientific Legitimacy: Misrepresented Science in the Direct-to-Consumer Genetic Testing Marketplace" (*Trends in Genetics,* 2010).

Coalition on the Public Understanding of Science

1444 I ("Eye") Street, NW, Suite 200
Washington, DC 20005
Phone: (202) 628–1500
Fax: (202) 628–1509
E-mail: copus@aibs.org
URL: http://www.copusproject.org/

"There appears to be broad consensus that the entire scientific ethos is under assault by uninformed or intentional forces. Science is poorly understood and often misrepresented to the public, the news media, and to political decision-makers." This statement constitutes one part of the original document on which the Coalition on the Public Understanding of Science (COPUS) was founded in 2007. That document grew out of a workshop funded by the National Science Foundation (NSF) and attended by a diverse group of participants, primarily from the biological and geological sciences, concerned about what they saw as the deterioration in understanding of and appreciation for the scientific enterprise in the United States. Attendees at the workshop agreed that a concerted effort was needed to better inform the general public and policy makers about the ways in which science functions, the nature of the results produced by scientific research, and the ways in which that information can be used for the betterment of society. Workshop participants drew up and adopted a "concept paper" that summarized the group's views as to why a group such as COPUS was needed, what is being done to deal with the current problem, what is missing or is not being done that should be done, what a new organization (COPUS) should focus on doing, and what the specific activities of COPUS might be. (The concept paper is available online at http://www.copusproject.org/concept-new.html.)

Some of the activities proposed for COPUS in the original concept paper included the following:

- Establishing a think tank to develop strategies for improving public understanding of science.

- Clarifying a specific message to be delivered to a range of audiences, including scientists and the technically literate, students at the K-16 level, members of the media, the general public, decision makers, and members of the business community.

- Preparing clearly worded explanations of the scientific process and the nature of science for use in a variety of settings, such as science textbooks.

- Developing an Understanding Science web site that could be based on the existing Understanding Evolution web site.

- Providing scientists and proponents of science with the tools and information to better explain the nature of science and the scientific process, as well as the potential and limitations of science.

- Empowering 1 million scientists by providing them with the confidence and tools to respond to the media, testify before legislative committees and other decision-making groups, addressing school boards, etc.

- Supporting broadcast facilities to produce and distribute radio and video productions, ads, and announcements explaining the nature of science.

The organization's first major activity was the Year of Science, celebrated in 2009. That event involved a host of activities sponsored by COPUS and conducted at communities of all sizes and composition throughout the United States. A specific theme was selected for each month of the year, including Process and Nature of Science: Communication Science (January), Evolution (February), Physics and Technology (March), Energy Resources (April), and Sustainability and the Environment (May).

COPUS has also now developed a number of community-based activities to carry out its core objectives. These activities include the following:

- Science zines, which are similar to magazines focused on science, but are much smaller and simpler to produce. Ideas for developing, producing, and distributing science zines are available from the group's Small Science Collective at http://smallsciencezines.blogspot.com/.

- Citizen Science, a program that aims to involve individual citizens in scientific projects. Suggestions for developing and carrying out such products can be found at the organization's web site, SciStarter, http://scistarter.com/index.html.

- Science Café, similar in concept to an old-fashioned kaffee klatsch, at which individuals with common interests meet in an information setting to listen to presentations, discuss issues of importance, and otherwise share their views on topics.

- Science festivals are somewhat more ambitious projects modeled to some extent on the now very popular national USA Science & Engineering Festival, held annually since 2010 at the National Mall in Washington, D.C. Such festivals are an opportunity for individuals and groups to exhibit their own projects and/or activities in science in a format roughly comparable to that of familiar school science fairs.

As of late 2013, the COPUS network consisted of 945 diverse groups ranging from A. C. Moseley High School in Lynn Haven, Florida, and the Alachua County (Florida) Office of Waste Alternatives to the Yale Peabody Museum of Natural History and Yuri's Night, self-labeled as "The World's Space Party." A key segment of the organization is the COPUS Core, a group of about a dozen individuals who first got together at a meeting at the University of California at Berkeley in March 2010 to talk about and lay out possible future directions for the organization. Each member of the Core has expertise in a particular area which they use to design and promote specific activities for the group. Some members of the Core include Lee Allison, State Geologist and Director, Arizona Geological Survey; Larry Bock, founder of the USA Science & Engineering Festival; Darlene Cavalier, writer for *Discover Magazine* and blogger under the name of The Science Cheerleader; Jennifer Collins, of the Consortium for Ocean Leadership, Deep Earth Academy; and Natalie Kuldell, instructor in the Department of Biological Engineering, Massachusetts Institute of Technology.

Teresa Stanton Collett

Same-sex marriage has been an issue of major interest in the United States and many other countries of the world since at least the end of the 20th century. During that period of time, public attitudes in most nations, including the United States, has gone from strong disapproval to weak to strong majority approval in many nations. As of 2014, 15 nations and 13 U.S. states and the District of Columbia permitted same-sex couples to marry.

The debate over same-sex marriage has been a complex and contentious one, involving issues from religion, politics, economics, law, social theory, psychology, and other fields of human scholarship. That debate involves some issues for which there probably is no clear right or wrong; for example, one can hardly say that one's religious view on the topic can be called "correct" or "false." Other issues, however, have a strong scientific basis; it is possible to say whether the scientific information presented on the topic is generally accepted by specialists in the field or not.

One feature of the ongoing debate over same-sex marriage, in fact, has to do with the accuracy with which spokespersons on both sides of the discussion have presented the best available scientific knowledge on a topic. For example, one well-known opponent of same-sex marriage is Teresa Stanton Collett, Professor of Law, at the University of St. Thomas, in St. Paul, Minnesota. Collett has taken the position on a number of occasions that same-sex marriage should be discouraged because it causes harm to children who may be involved in a same-sex family. In an article for the *William Mitchell Law Review* in 2007, for example, Collett wrote that "children best flourish when raised by their biological mother and father who are united in marriage," and, later, that

> [t]here is a growing consensus in the social science literature that clearly establishes that children do best when they are raised by both biological parents who are married to each other.

This argument would have to be taken seriously if it truly represents the best judgment of specialists in the field: psychologists, sociologists, child welfare agents, and the like. But such has hardly been the case. Indeed, one review of the argument about child welfare in same-sex households points out that

> [e]very mainstream health and child welfare organization—including the Child Welfare League of America, the American Academy of Pediatrics, and the American Psychological Association—has issued policies opposing restrictions on lesbian and gay parenting. (Courtney G. Joslin, "Searching for Harm: Same-Sex Marriage and the Well-Being of Children." *Harvard Civil Rights-Civil Liberties Law Review*, 46[81])

Also, a comprehensive study of more than 80 research studies, books, and articles by experts in the field conducted in 2013 concluded that children raised in same-sex homes are not harmed by that type of family and, in fact, such households actually strengthen families and the children raised in them. The study, which appeared in the March 20, 2013, online version of the journal *Pediatrics*, was cited as being influential in the decision announced in the same issue by the American Academy of Pediatrics to support legalization of same-sex marriage. The same position was taken by U.S. Supreme Court Justice Anthony Kennedy in the Court's 2012 decision to invalidate the U.S. Defense of Marriage Act (DOMA). Kennedy argued in that decision that the act actually caused harm to children of same-sex couples by opening them up to unfounded criticism of the type of household in which they grow up.

The question arises, then, as to why a highly respected expert in the field such as Collett would apparently attempt to buttress her argument about a social and political issue while ignoring the best available scientific evidence on a topic. She is hardly alone, however, in following a practice common when politicians deal with controversial issues having a basis in science that involves ignoring or misrepresenting (intentionally or unintentionally) what specialists in the field have to say on the topic.

Teresa Stanton Collett attended the University of Oklahoma, from which she received her bachelor of arts in letters degree in 1977. She received her JD degree, also from the University of Oklahoma in 1986. (Her official biography provides no information on her activities between 1977 and 1986.) Collett worked for three years as an attorney at Crowe and Dunlevy, in Oklahoma City, before taking a position as visiting assistant professor of law at the University of Oklahoma from 1989 to 1990. She then accepted an appointment as professor of law at the South Texas College of Law in Houston, where she remained until 2003. Collett then moved to the University of St. Thomas School of Law, where she has remained ever since. She has also served as visiting professor of law during the summer term at the University of Houston Law Center (1992), University of Texas School of Law (1995), and Washington University School of Law (1996). She also taught at Notre Dame University during the period August 1995 through July 1996 and August 1996 through July 1997.

Throughout her professional career, Collett has been active in speaking and writing about and acting on behalf of issues about which she is especially concerned, including the anti-abortion movement and the anti-same-sex marriage movement. She has served as special attorney general to the states of Oklahoma and Kansas, and has advised the attorneys general of other states, especially on issues of reproductive rights and abortion. She has offered *amici curiae* briefs on behalf of former Surgeon General C. Everett Koop, in opposition to the state of Oregon's assisted suicide law; on behalf of Tim Pawlenty and John Hoeven, governors of the states of Minnesota and North Dakota in support of their parental notification (of abortion) laws; and on behalf of the state of Oklahoma in support of its abortion liability and parental notification laws. She has also been involved in a host of other cases involving abortion and same-sex marriage on behalf of organizations such as the Catholic Medical Association, the Texas Eagle Forum, the Christian Medical Association, and the American Association of Pro-Life

Obstetricians/Gynecologists. Collett ran for the U.S. House of Representatives from Minnesota's Fourth District in 2010, losing by a margin of 59 percent to 35 percent to Representative Betty McCollum.

Federation of American Scientists

1725 DeSales St., NW, Suite 600
Washington, DC 20036–4413
Phone: (202) 546–3300
Fax: (202) 675–1010
E-mail: fas@fas.org
URL: http://www.fas.org

Most people are familiar with one result of the Manhattan Project, created during World War II to develop a nuclear weapon to be used against the Japanese. Far fewer people are aware of a second consequence of that project, the creation of a new appreciation for the burgeoning power of scientific research, as expressed in the development of nuclear weapons, and its potential significance in the direction of political and policy decisions in nations around the world. Scientists involved in the Manhattan Project were faced with the challenge of considering the unheard of power that they had released and the potential consequences it had for the human race.

Even before the war ended, many of these scientists began to form groups to talk about these consequences and ask what their role, both as scientists and possibly as politicians, was to be in the postwar years. Such groups were formed at most Manhattan research centers, including the Metallurgical Laboratory at the University of Chicago, the Clinton Laboratory at Oak Ridge, the Los Alamos Laboratory, and the Substitute Alloy Materials Lab at Columbia University. Wartime security was so tight that these groups had essentially no contact with each other. Indeed, members of two such groups at Oak Ridge, the Atomic Production Scientists of Oak Ridge and the Association of Oak Ridge Scientists, were unaware of each other's existence.

As the war came to an end, secrecy was relaxed, and groups from the various research centers learned about each other's work and decided to call a national conference. That conference was held in late 1945, and on November 1 of that year, representatives of the various groups announced the formation of the Federation of Atomic Scientists (FAS). (Only two months later, it changed its name to its current form, the Federation of American Scientists.) The first objectives of that group were to ensure transparency in future nuclear research and to work for international control of nuclear power. These goals were driven by their realization that the potential to build nuclear weapons would inevitably spread throughout the developed world, creating a host of problems that only open discussion among all participants could solve.

Although many of the initial issues troubling atomic researchers were eventually resolved, not always to the group's satisfaction, the FAS continued to focus on nuclear issues until the 1960s. At that point, the organization decided to expand its agenda to include a number of other social issues with a strong scientific component. In the process, FAS made a concerted effort to broaden the composition of its membership, reaching out to researchers from all fields of science who were concerned with the social implications of their work. The group's agenda today continues to reflect that eclectic emphasis. Today, seven major themes dominate the work of the Federation: the Nuclear Information Project, government secrecy, arms trade biosecurity, learning technologies, international science partnerships, and emerging technologies.

The Nuclear Information Project carries on the oldest of the FAS objectives, developing, collating, and distributing information about nuclear weapons and commercial nuclear power facilities to the general public and to those individuals who are in policy-making positions in government. A key component of the project is a collection of publications, many by the FAS itself and others by like-minded organizations, on topics such as nuclear missiles, nuclear waste, nuclear policy, power plants, and national policy. The project also maintains an archive of current news articles from print and electronic media around the nation.

The government secrecy project also reflects one of the early emphases of the organization, an effort to provide maximum transparency for a field that has a tendency to hold in reserve a great deal of the new information gained on the subject of nuclear power. The secrecy project works to promote public accessibility to information about nuclear power that does not have to be kept secret for national security reasons and to publish, where necessary, government and other documents that are otherwise unavailable to the general public and/or the research community.

The Arms Sales Monitoring Project (ASMP) focuses on another area in which secrecy tends to be the standard by which governments, corporations, and individuals operate, the cross-border purchase and sale of all types of weapons. The goal of ASMP is to collect information about otherwise clandestine arms trades and make that information available to the general public, journalists, government officials, and other policy makers. The biosecurity project has taken on the problem of chemical and biological weapons, different in form from, but similar in their potential ultimate consequences to, nuclear power. The FAS provides an Internet resource on the nature of such weapons, their potential use by nations, and consequences that can be expected from such use.

FAS created the Learning Technologies Program (LTP) in 2001 to give consideration to the rapidly growing and expanding forms of technology that can be used both within traditional classrooms and outside of them. The project attempts to find out about such technologies and to provide information about them to educators, students, and policy makers. The project explores the potential of any and all means—ranging from video games to zombies—as potential tools in the teaching and learning of abstract topics such as molecular biology.

The International Science Partnership was created in 2012 for the purpose of bringing scientists from a variety of nations together to work on issues of common interest. The pilot project in the Partnership is one between scientists from the United States and Yemen, working together on water and energy issues of common interest in the two nations. The Emerging Technologies and High-End Threats (ET/HT) Project is yet another activity that reflects the early goals and objectives of the FAS. The project asks the question as to what new types of technology might be able to change the nature of national and international conflict in as profound a manner as did the invention of nuclear weapons at the end of World War II. Some of the potential weapon systems that are being studied in this context are those that make use of nanotechnology, robotics, electromagnetic radiation, lasers, genetics, meteorology, and cyber activities.

Sir Francis Galton (1822–1911)

Important scientific discoveries sometimes have momentous impact on social, political, economic, moral, and other aspects of society. Chapter 1 of this book discusses, for example, the social, political, and religious disruptions caused by Galileo's discoveries of the moons of Jupiter and other telescopic findings. A more recent example of this phenomenon is the appearance of the theory of eugenics in the last quarter of the 19th century.

The term *eugenics* refers to the theory that human civilization can be improved if efforts are made to control the breeding choices made by individual men and women. That theory has existed since the Greek and Roman civilizations when governments were actively engaged in deciding which newborn children should be allowed to survive because they would be assets to the state, and which should be put to death because they would be only burdens to society. The modern theory of eugenics, however, owes its existence to the work of Sir Francis Galton.

Galton was half cousin to Charles Darwin, who had proposed his theory of natural selection in his 1859 book, *On the Origin of Species.* Darwin had argued that individuals best adapted to their environment lived longer than their weaker peers, thus increasing the probability of their passing on their genetic traits to their descendants. From a long-term view, natural selection was the mechanism, Darwin said, by which species survived, adapted to new environmental conditions, and became stronger overall.

Galton was convinced that Darwin's theory applied to humans as much as it did to other animals and plants. The problem he envisioned, however, was that the beneficence of human civilization interfered with the normal operation of natural selection. In Galton's world, a variety of state activities, such as public hospitals and orphanages, had the result of keeping alive individuals who were "unfit," allowing them to pass their genes on to the next generation, and producing a weaker race overall. He proposed that states should take quite a reverse approach to dealing with the "less fit" among them, but making sure, insofar as possible, that those who were less healthy, less intelligent, or morally unqualified be prevented from having children. One of the most obvious ways of doing that was to sterilize men and women who could be classified into one of these categories.

During the first third of the 20th century, the eugenics movement became very popular in a number of developed nations, including the United States. Indiana became the first state in 1907 to pass a sterilization law that allowed the government to identify men and women who were "unfit" to have children and to have those individuals sterilized without their consent. Over the years, 30 more states adopted similar laws, with South Carolina being the last of these, in 1935. At about the same time, the National Socialist Party (Nazis) in Germany adopted a similar set of laws aimed to the forceful elimination of the Jewish race. By the late 1960s, the eugenic movement had largely died out in the United States, with the Supreme

Court's overturning of Virginia's Racial Integrity Act of 1924 in the case of *Loving v. Virginia*. By 1972, the last of the state sterilization laws had been revoked.

Francis Galton was born in Birmingham, England, on February 16, 1822, to a highly respected family that numbered among its members the great English physician Erasmus Darwin. Galton was by all accounts a child prodigy who taught himself to read not only English, but also Greek and Latin by the age of five. Galton enrolled at King Edward's School in Birmingham, but was bored by the curriculum and withdrew at the age of 16. He then decided to pursue a medical career, matriculating at the Birmingham General Hospital and King's College, London Medical School. Upon completion of his studies, he studied mathematics at Trinity College, Cambridge, from 1840 to 1844.

When his father died in 1844, Galton came into a considerable fortune, and no longer had to consider a possible career; he could do whatever he wanted with his life. He chose to spend his time traveling and pursuing a variety of scientific interests in meteorology, statistics, psychology, biology, and criminology. For example, he constructed the first weather maps of the type that still appear in the daily newspaper, developed mathematical procedures for the analysis of statistical regression and correlation, constructed theories of the transmission of genetic characteristics, and suggested a system of using fingerprints for the identification of criminals.

During his lifetime, Galton was honored with a number of prestigious awards, including the Silver Medal of the French Geographical Society, the Gold Medal and the Darwin Medal of the Royal Society, the Huxley Medal of the Anthropological Institute, and the Darwin-Wallace Medal of the Linnean Society of London. He was named an Officier de l'Instruction Publique by the French government in 1891, elected an honorary fellow of Trinity College, Cambridge, in 1902, and knighted by King Edward VII in 1909. The Galton Chair of Genetics (originally the Galton Chair of Eugenics) at University College

London is named in his honor. Galton was the author of two very influential books in the fields of genetics and eugenics, *Hereditary Genius* (1869) and *Inquiries into Human Faculty and Its Development* (1883). In the hope of making his ideas about eugenics more widely known, he also wrote a novel describing a society based on the principles of eugenics, which he called *Kantsaywhere*. He was unable to find a publisher for the book, however, with the result that it remained largely unknown until 2011, when UCL, the publishing arm of University College London, made the book available online at http://www.ucl.ac.uk/library/special-coll/ksw.shtml.

Galton died in Haslemere, Surrey, England, on January 17, 1911, two weeks before the opening of the First International Congress of Eugenics in London.

Government Accountability Project

1612 K St., NW, Suite #1100
Washington, DC 20006
Phone: (202) 457–0034
E-mail: info@whistleblower.org
URL: http://www.whistleblower.org

The Governmental Accountability Project (GAP) is a nonprofit 501(c)(3) organization founded in 1977 to protect whistle-blowers who speak out regarding corporate and governmental misuse of their powers and authorities. The organization is also committed to informing employees and the general citizenry of its right of free speech in addressing issues of concern not only to groups of employees, but also to the nation as a whole. GAP was created originally in 1975 as a program of the Institute for Policy Studies (IPS) to deal specifically with whistle-blower issues (a topic with which IPS continues to be concerned). It was chartered as a separation organization two years later, although it then continued a close relationship with its parent organization, IPS. GAP currently has an operating annual budget of about $2.5 million, which comes from about

10,000 individual donors, as well as a number of charitable organizations, such as the Rockefeller Family Fund and the Open Society Foundation.

The primary focus of much of GAP's early work was the nuclear power industry. A number of employees of that industry in the late 1970s and early 1980s were concerned about safety issues related to the construction and operation of nuclear power plants and the storage of and risks posed by nuclear wastes. Those employees faced risks of losing their jobs or other punishments, however, to the extent that they chose to speak out about those issues. IPS and GAP stepped into this dispute on behalf of concerned employees, worked to make it possible for them to express their concerns and views, and helped to protect them against recriminations and rebukes by their employers.

Nuclear power generation remains one of GAP's primary programmatic concerns, now listed under the general rubric of Environment. This program area also includes controversial areas such as the operation of governmental environmental agencies and climate change science and issues. Other program areas include food integrity, corporate and financial responsibility, government employees, national security and human rights, international affairs, legislation, litigation, and public health. Some examples in each of these areas are as follows:

Environment: GAP has worked to expose efforts by government officials to influence and/or alter scientific reports on climate change used to make public policy and/or to inform the general public on this contentious issue.

Food Integrity: The GAP Food Integrity Campaign is designed to ensure the safety of food made available to the American public by insisting that accurate information is available about food protection practices.

Corporate and Financial Responsibility: This program is aimed at publicizing wrongdoings by corporate and financial institutions and working to protect whistle-blowers who provide information about illegal and unfair practices by such institutions.

Government Employees: One of GAP's major activities is defending and supporting whistle-blowers in federal government when they speak out about illegal or questionable actions by their employees. This program tends to overlap some of the other program areas listed here, since many whistle-blowers raise issues within these specific topics.

National Security and Human Rights: This program deals with some of the most difficult freedom-of-speech areas, since whistle-blowers may choose to speak out about areas of high security, where transparency is not encouraged or even possible. Some specific issues with which GAP has dealt include the ways in which governmental agencies have dealt with disasters, politically motivated discrimination, secrecy of policies and practices, surveillance of innocent Americans, and the use of torture by governmental agencies.

International Affairs: The international program area deals with agencies and activities that go beyond national boundaries, such as the politics and activities of agencies such as the International Monetary Fund and the World Bank.

Legislation: In a somewhat different vein, GAP works for the adoption of laws and regulations that provide some measure of protection for individual employees who raise questions about or speak out about questionable corporate or governmental practices.

Litigation: The organization also works concretely to work in the defense of individual whistle-blowers who have been harassed or punished because of their actions in drawing attention to company or agency policies and practices.

Public Health: One of the areas that impact the general public very directly is in the inactions or misactions of corporate and governmental entities involving the products and practices of public health. GAP works to reveal these deficiencies and to support and protect employees who speak out against such activities.

GAP has created a number of avenues through which to carry out its work. For example, it has developed specific tools for workers in specific fields to use in their efforts to learn more about and to speak out about questionable activities in their field of employment. An example of such a tool is the organization's current Banking Know Your Rights Campaign, which consists of a web site, a hotline, pamphlets, and a handbook designed to help financial workers decide whether practices they observe are legal or not and what to do if they decide to speak out against such practices. GAP also sponsors petition programs that provide individual citizens with a means by which to express their views on controversial issues. An example of such a campaign was the petition to stop the use of certain toxic dispersants after the BP oil spill in the Gulf of Mexico in 2010.

GAP has also become part of or spearheaded a number of coalitions dedicated to dealing with specific topics. An example of such a coalition is the Make It Safe Coalition, consisting of more than 50 organizations whose goal is to make it safer for individual whistle-blowers to speak out about issues they regard as important to American health and safety. Over the past few years, GAP has also been operating the American Whistleblower Tour, which visits about two dozen college and university campuses to discuss the role of whistle-blowing and the programs that GAP sponsors to support the work of whistle-blowers.

Edward R. Korman (1942–)

Korman is a U.S. district judge for the Eastern District of New York. Since 2008, he has also been sitting by designation on the Ninth Circuit Court of Appeals in San Francisco. (The term *by designation* means as a visiting judge.) Korman is probably best known to the general public for his decisions relating to administrative actions by the U.S. Food and Drug Administration (FDA) with regard to the dispensing of the contraceptive drug (levonorgestrel). Korman has been presiding judge in cases relating to the use of Plan B since January 2005, when proponents of the plan filed suit to have the drug made available on an over-the-counter (OTC) basis for women of all ages. In the eight years during which Korman listened to arguments from the government suggesting that Plan B should be limited to women, first as a prescription drug only, and later for women of a certain age only (an age that continued to change over the years), he appeared to become increasingly frustrated at the government's unwillingness to accept generally accepted scientific opinion (including that from FDA's own panel of experts) that Plan B was safe and effective for women of all ages. His final ruling in 2013 spoke clearly and unambiguously about the government's diversionary tactics and ordered the FDA to make Plan B available to women of all ages without prescription within 30 days of his ruling. The FDA actually acted within days to do so.

Edward Robert Korman was born in New York City on October 25, 1942. He received his BA from Brooklyn College in 1963, his LLB from Brooklyn Law School in 1966, and his LLM from New York University in 1971. He served as law clerk for Kenneth B. Keating of the New York Court of Appeals from 1966 to 1968 before joining the New York City law firm of Paul Weiss Rifkind Wharton & Garrison. He left Paul Weiss in 1970 to become an assistant United States attorney for the Eastern District of New York (1970–1972) and then took the position of assistant to the Solicitor General of the United States from 1972 to 1974. He joined the U.S. Attorney's Office for the Eastern District of New York once more in 1974, where he served as chief assistant United States attorney until 1978 and United States attorney until 1982.

In 1982, Korman left public service to become partner at the law firm of Stroock & Stroock & Lavan, where he remained until 1985. During this period he was also professor of law at Brooklyn Law School and a member of the Temporary Commission of Investigation of the State of New York and Chairman of the Mayor's Committee on New York City Marshals. On October 2, 1985, he was nominated to be a judge in the U.S. District Court for the Eastern District of New York, a post for which he was confirmed by the U.S. Senate on November 1, 1985. Korman has been honored with the Federal Bar Council's Learned Hand Medal for Excellence in Federal Jurisprudence and the New York County Lawyers' Association's Edward Weinfeld Award for Distinguished Contributions to the Administration of Justice. He is the author of a 2006 essay, "Rewriting the Holocaust History of the Swiss Banks: A Growing Scandal," published in *Holocaust Restitution: Perspectives on the Litigation and Its Legacy*, edited by Michael Bazyler and Roger P. Alford (New York University Press) and a biographical essay on Kenneth Keating, published in *The Judges of the New York Court of Appeals, A Biographical History*, edited by Albert M. Rosenthal (Fordham University Press, 2007).

Philipp Lenard (1862–1947)

One of the most important reasons for the loss of World War II by Germany and Italy was the exodus from those two countries of some of their best qualified and most famous scientists. For example, many of the researchers responsible for developing the theory on which nuclear weapons are based and the technology from which they were eventually produced left the two nations during the 1930s, largely because of fears for their lives. This exodus simultaneously deprived Germany of the intellectual resources needed to build a nuclear weapon and made available those resources to the Allied nations. Leading researchers in a number of other fields of science also decided to flee to Great Britain, the United States, and other nations.

The flight of so many scientists from Germany (and, to a lesser extent, Italy) occurred because of the strongly anti-Semitic policies of the Nazi party in Germany and its fascist counter-part in Italy. By the early 1930s, new German laws had begun to exclude Jews from all professions, including all teaching and research positions, and the first programs for the extermination of Jews were already underway.

An important element in the anti-Jewish campaign in Germany was an attack within the scientific community on the so-called Jewish science. This attack was based on a philosophy that the fundamental ideas of modern physics then being taught and practiced by leading German physicists (such as Albert Einstein) were based on false theories, such as notions of relativity and quantum mechanics. Supporters of the Nazi party argued that loyal citizens of the nation had an obligation to develop and teach a new, different, more "pure" form of physics, which came to be called *Deutsche Physik* (German physics, or Aryan physics). Perhaps the most enthusiastic proponent of this view was the renowned physicist Philipp Lenard.

Lenard became a member of the National Socialist (Nazi) party when it was still a young and growing movement and rapidly became one of the most trusted advisors to Adolf Hitler. As early as 1924, he had described Hitler in one of his lectures as a "true philosopher with a clear mind." Over the years, Hitler often turned to Lenard for his advice on scientific matters, respecting his views because he was one of the few leaders in the field who had not abandoned the New Reich after it had reached power.

Lenard's views on the proper approach to the learning and teaching of physics are probably best expressed in his four-volume work, *Deutsche Physik* (*German Physics*), published in 1936 and 1937. In this work, Lenard focused on the concepts of classical physics, as it was understood in the late 19th century. The book largely ignored the scientific revolution that took place in physics during the first decade of the 20th century and emphasized experimental results and practical applications.

Lenard clearly expressed his views about German physics in the introduction to *Deutsche Physik*. Most observers have argued, he wrote, that "Science is and remains international." Such a view, he went on, is false. Instead, "Science, like every other human product, is racial and conditioned by blood." Thus, the Nazi party had an obligation to cleanse the nation of false and misleading ideas in science and replace them with a more "pure" approach to science based in experimentation and reality-based facts, rather than speculative theories.

Philipp Eduard Anton von Lenard was born on June 7, 1862, in Pressburg, then part of the Austro-Hungarian Empire. He received his earliest education at home before enrolling at the cathedral school at Pressburg at the age of nine. He later transferred to the Pressburg Realschule, where he completed his secondary education. Lenard was fascinated by science from his earliest days, and he eagerly looked forward to continuing his education at the university level. His father, however, had different plans, hoping and expecting that his son would join him in the wine business. Eventually, the senior Lenard agreed to allow his son to attend courses at two technical schools, the Technische Hochsculen in Vienna and Budapest. At the conclusion of this relatively unsatisfactory (for Lenard) experience, he returned to Pressburg to work at his father's wine shop.

It was not until 1883 that Lenard had saved up enough money to begin the university education for which he had so hoped. He traveled to Heidelberg, where he spent four semesters largely under the tutelage of the famous Robert Bunsen, whom he later described as his "secret object of worship." He then continued his studies at the University of Berlin, where he began his doctoral studies on the oscillation of water droplets. For this research, he received his doctorate, summa cum laude, from Heidelberg in 1886.

Over the next two decades, Lenard moved from one university to another, at the City and Guilds Electromagnetic and Engineering Laboratories of the London Central Institution; the University of Breslau, in what is now Poland; at the University of Bonn, where he worked under Heinrich Hertz; once more at the University of Heidelberg, from 1896 to 1898; and at the University of Kiel, from 1898 to 1907. In 1907, he finally returned once more to Heidelberg, as professor of physics and director of the radiology laboratory (which was renamed the Philipp Lenard Laboratory in his honor in 1935). By that time, he had already won the 1905 Nobel Prize in Physics, for his research on the properties of cathode rays. He continued to work at Heidelberg until his retirement in 1931, after which he remained at the university with emeritus status. He was finally expelled from this post by Allied occupation forces in 1945. He died in Messelhausen, Germany, on May 20, 1947. In spite of his strong and venomous political views, Lenard was widely respected within the field of physics, having been honored with the Rumford Medal of the Royal Society of Great Britain, the Matteucci Medal of the Italian Society of Sciences, and the Franklin Medal of the U.S. Franklin Institute.

John Marburger (1941–2011)

Marburger served as director of the Office of Science and Technology Policy (OSTP) under President George W. Bush from 2001 to 2009. He served longer in that position than did any other person to hold that title. His long tenure was characterized, however, by an ongoing struggle with critics of the Bush administration's positions on a variety of important scientific issues. Throughout that period, however, Marburger remained loyal to Bush, mounting well-thought-out and sound counterarguments to critics who objected to what they perceived as the President's politicization of a number of scientific problems facing the nation.

Marburger's tenure did not begin fortuitously. He was not even confirmed by the U.S. Senate until October 2001, nearly 20 months after Bush had been inaugurated and a month after he had been nominated by the President. In the meanwhile, the President had already laid out his administration's position on a number of important scientific issues. As Sherwood Boehlert (R-N.Y.), chair of the House Committee on Science, has noted, the challenge Marburger faced was "serving a president who didn't really want much scientific advice, and who let politics dictate the direction of his science policy." In addition, Marburger had to deal with the problem of "being someone who had earned the respect of his scientific colleagues while having to be identified with policies that were not science-based."

Whatever personal views he may have held on the controversial issues he had to deal with (and he had said that he would always keep his personal views private), he always saw it as his job to explain and defend the administration's position on those issues. For example, when the Union of Concerned Scientists (UCS) released its monumental report on the treatment of scientific issues by the Bush administration, *Scientific Integrity in Policymaking: An Investigation into the Bush Administration's Misuse of Science*, in 2004, Marburger prepared a detailed, point-by-point response which prompted the UCS to fine-tune and in turn modify its report.

John Harmen Marburger III was born in Staten Island, New York, to Virginia Smith and John H. Marburger II on February 8, 1941. He earned a bachelor of arts degree in physics from Princeton University in 1962 and a PhD in applied physics from Stanford University in 1967. After graduation, he took a position as professor of physics and electrical engineering at the University of Southern California (USC), where he remained until 1980. During his tenure at USC, he served as chair of the physics department and dean of the College of Letters, Arts, and Science. He also cofounded the Center for Laser Studies at USC.

In 1980, Marburger accepted appointment as president of the State University of New York at Stony Brook (SUNY-SB), a post he held until 1994. He then returned to research and teaching at SUNY-SB until 1998, when he was appointed president of Brookhaven Science Associates, an organization created to bid for the right to manage and operated the Brookhaven National Laboratory, a bid that the company eventually won.

Marburger served as director at Brookhaven until his appointment by President Bush in 2001.

After leaving government service in 2009, Marburger returned to SUNY-SB as a member of the faculty and as vice president for research. By that time, however, he had become ill with non-Hodgkin's lymphoma, a condition from which he eventually died on July 28, 2011, at his home in Port Jefferson, New York. One of the accomplishments for which he may be best remembered within the scientific community was his advocacy for a new field of science, sometimes called the science of science policy. The field is concerned with a rigorous examination of the factors involved in the development and implementation of scientific policy on international, national, and regional levels. In 2011, he was coeditor of a book on the topic, *The Science of Science Policy: A Handbook*. His textbook on quantum mechanics, *Constructing Reality: Quantum Theory and Particle Physics*, was published posthumously in late 2011.

National Academy of Sciences

500 Fifth St., NW
Washington, DC 20001
Phone: (202) 334–2000
E-mail: http://www.nasonline.org/about-nas/contact-us/
URL: http://www.nasonline.org/

The National Academy of Sciences (NAS) was created by act of the U.S. Congress on February 20, 1863, in the midst of the Civil War. The organization was created under the impetus of a group of professional scientists, most of whom were from Cambridge, Massachusetts. The purpose of the NAS was to establish an organization that could provide the U.S. government with independent, objective advice on scientific issues of concern to the United States. The organization is a nonprofit organization that consists of the most eminent scientists in the country, who are elected to the body by their peers. Membership has grown from the initial 50 founding fathers (and they were all men) to more than 2,200 members today, along with 400 foreign associate members. Under current rules, no more than 84 new members and 21 new associates may be elected in any one year. Election to the NAS is one of the highest honors that can come to an American scientist or to prominent scientists from around the world.

Over the years, three important branches have been added to the original organization, the National Research Council (NRC), the National Academy of Engineering (NAE), and the Institute of Medicine (IOM). The NRC was created in 1916 by President Woodrow Wilson, who felt that the wartime needs of the United States could be aided by the recruitment of a large body of well-trained research to work on military problems. After the war, Wilson issued an executive order allowing the NRC to consider its operation into the postwar era. The NAE was created in 1964 when Congress amended the original NAS founding act to create an organization focused more on practical applications of science. The NAE currently has more than 2,000 members and foreign associates. The establishing act was extended once more with the creation of the IOM, which also has more than 2,000 members and foreign associates. As its name suggests, the IOM focuses on the nation's health and medical issues. The collective organization of the NAS, NAE, IOM, and NRC is now known collectively as The National Academies.

The work of the NAS is distributed among eight ongoing programs and additional specialized programs. The former group consists of the following programs:

- Koshland Science Museum, an institution located in Washington, D.C., which presents exhibits and events dealing with scientific issues that people encounter in their everyday lives

- Cultural Programs, a series of lectures, concerts, films, and art exhibitions that explore the intersections among the arts, science, medicine, engineering, and popular culture

- Sackler Colloquia, discussions among experts in the field on topics that cross a variety of scientific and nonscientific disciplines

- Kavli Frontiers of Science, symposia at which young scientists exchange views on important cross-disciplinary, innovative issues

- Distinctive Voices, public meetings that present and explain important new ideas and developments in the sciences, with their implications for the everyday lives of humans

- Sackler Forum, a series of lectures and discussions by representatives from the United Kingdom and the United States about important international science-based issues

- Keck Futures Initiative, a program to support and reward breakthrough research at the intersection of science, engineering, and medicine

- Science and Entertainment Exchange, a program to promote the interaction between scientists and representatives of the entertainment business to find ways of making science more understandable and interesting to the average person.

Four specialized programs as of 2014 dealt with the Gulf of Mexico, resources on evolution, climate change, and science education. In each of these areas, NAS promotes and encourages discussion about issues of importance to the United States in the 21st century, along with providing expertise and resources that help with understanding and working on these issues.

One of NAS' most important contributions both to the scientific community and to the general public is its publication arm, which produces three important journals and authoritative books on the most important scientific and health problems of the day. The three journals are the *Proceedings of the National Academy of Sciences* (*PNSA*), which publishes research reports, commentaries, reviews, perspectives, and colloquium papers on topics in the physical, biological, and social sciences; *Issues in Science and Technology*, which focuses on policy implications of developments in science and technology; and *Biographical Memoirs*, which carries obituaries of members. Many of the books published by the National Academies Press are now classics in their field on topics such as Climate and Social Stress: Implications for Security Analysis; Disaster Risk Management in an Age of Climate Change; Seasonal-to-Decadal Predictions of Arctic Sea Ice: Challenges and Strategies; Assessing Risks to Endangered and Threatened Species from Pesticides; Antibiotic Resistance: Implications for Global Health and Novel Intervention Strategies; Research for a Future in Space: The Role of Life and Physical Sciences; Marijuana as Medicine?—The Science Beyond the Controversy; and Preparing for the Future of HIV/AIDS in Africa—A Shared Responsibility.

In addition to its ambitious print program, the NAS provides a full line of electronic resources on its web site, including news articles on topics of interest in science and its relationship to society, podcasts, social media connections, a number of mobile apps, and a video gallery.

National Center for Science Education

420 40th St., Suite 2
Oakland, CA 94609–2688
Phone: (510) 601–7203
Phone Toll Free: (800) 290–6006
Fax: (510) 601–7204
E-mail: info@ncse.com
URL: http://ncse.com/

The National Center for Science Education (NCSE) is an organization devoted to providing information to educators, students, policy makers, and the general public about the controversy concerning the teaching of evolution and global climate change. The organization was founded in 1981 by a group of concerned citizens who had become aware of and troubled by efforts in their home states to pass laws that would dilute the teaching of evolution in biology classes at the precollege level. The group was incorporated in 1983 and was able to establish a national office in Washington, D.C., in 1986, largely as a result of a grant from the Carnegie Foundation. Dr. Eugenie C. Scott, a physical anthropologist, was hired as director of the organization, a post that she continues to hold at the end of 2013.

NCSE is arguably the nation's single most valuable resource for the defense of the teaching of evolution in high school biology classes against the encroachments of efforts to add to that teaching information about creationism, creation science, intelligent design, and other interpretations of the changes that take place in organisms over the passage of time. One avenue through which the organization works is providing information and skill development to individuals and local groups directly, offering instruction in skills such as letter writing to governmental officials and other policy makers, testifying before a school board meeting, becoming better informed about local issues relating to the teaching of evolution, and networking with other groups with similar interests and objectives. NCSE also provides speakers and advisors who talk to and work with groups at the local and state level in dealing with specific state legislation. It also collects and distributes information on the current status of legislation in individual states on the teaching of evolution.

In 2012, NCSE expanded its traditional role beyond the evolution controversy to include the debate over global climate change. The organization saw many of the same issues and controversies that had developed over the teaching of evolution playing out in a new arena: the question as to how human activities affect global climate, if at all. Many of the resources that NCSE has developed for dealing with evolution deniers is now being put into practice for use against climate change deniers. The NCSE web page provides information on the status of scientific information about climate change, the nature of efforts to dispute the validity of that information, and the mechanisms that can be used by private citizens and small community groups to counter the actions of those who would introduce inaccurate information on the topic into the school curriculum.

Even before its official founding, members of the new NCSE had started publishing a journal dealing with issues related to the teaching of evolution, called the *Creation/Evolution Journal* (C/E J). A complete archival record of that journal is now available on the NCSE web site at http://ncse.com/media/cej. Simultaneously with C/E J, NCSE published an in-house publication, *NCSE Reports*. The two publications were merged in 1996 to produce *Reports of the National Center for Science Education* (RNCSE), which covers all aspects of the teaching of evolution, creationism, intelligent design, and global warming issues. The journal is available only online at http://reports.ncse.com/index.php/rncse. A second ongoing publication is the weekly electronic newsletter *Evolution Education and Climate Update*, which can also be retrieved at no cost on the NCSE web site at http://ncse.com/ncse-news-list.

In addition to these electronic resources, NCSE has produced a number of multimedia programs dealing with evolution and climate issues, including video programs on "Climate/Energy Literacy for the 21st Century," "Climate Change Denials and AFAs," "Science Swept under the Rug," and "Science Denial and Social Justice"; video presentations on topics such as evolution and science teaching; the evolution of creationism; controversies, scientific and otherwise; why we need to teach evolution; creationist tactics; and audio presentations on Hostile Climate, Inside the NGSS [Next Generation Science Standards], and Climate Change in Classrooms.

Plato (428/427–348/347 BCE)

The period between the sixth and third centuries BCE is sometimes thought of as the Golden Age of Greek science. It was during this era that a handful of thinkers developed a new system for studying the natural world, a system that was based on observation, to some extent, experimentation, and logical reasoning. This system, an early form of the modern scientific method, stood in contrast to traditional systems of knowledge, which were based primarily on the transmission of information from one generation to the next, with certain individuals recognized as indisputable authorities in a field.

To modern historians, this way of gaining knowledge was a brilliant and revolutionary advance in human culture. Many Greek scholars of that time, however, had a very different view of the rise of Ionian science, as the new philosophy was called. One of the greatest concerns expressed was that the democratization of knowledge posed a threat to the social order. Thus, leaders of the government and religious authorities were particularly concerned about the possible spread of the "new science" among the lower classes and common people. No one was more outspoken in this regard than the great philosopher, Plato.

Whatever the merits of Ionian science as a way of studying nature, Plato was convinced that giving just anyone the right to make observations and draw conclusions spelled the destruction of Greek society as he knew it. He felt so strongly that he taught that it was permissible for those in authority to invent a story that justified the structure of the existing society, essentially a religious tale in which the gods created humans in such a way that individuals were naturally different from each other, some better qualified to rule, the majority destined to serve the state and its masters. Plato acknowledged that this type of lie was impermissible in one's personal life, but it was not only allowed, but actually required, as part of a nation's social consciousness. He referred to that lie as The Noble Lie, because of the powerful effect it exerted on the stability of a society.

Given the power of the nobility and organized religion, it is hardly surprising that Plato's view of the "new science" eventually prevailed. At one point, about 100 years after Plato's birth, another philosopher, Epicurus, attempted to salvage the goals and methods of the new science. He was somewhat successful for a brief period of time, but by the end of the first century BCE his adherents had largely been silenced in Greece and expelled entirely from Rome. After a final noble effort to save the movement by the Roman poet Lucretius in his great poem *De rerum natura* (*On the Nature of Things*), the "new science" fell into oblivion, not to be revived for more than 15 centuries in the work of Galileo. Efforts by Plato and his sympathizers to control the direction of scientific research in its very earliest forms have turned out to be a harbinger of similar events that were to occur over and over again in the future interaction of politics and science in human culture.

Plato was born in either Athens or Aegina sometimes between 429 and 428 BCE. He came from a noble family descended on his father's side from kings of Athens and Messenia and on his mother's side, from the great Athenian lawmaker, Solon. Plato was inordinately fond of his family's history, and referred frequently to its most outstanding members in a number of his works. He was a student first of the most eminent of the Greek scholars of the time, Dicaearchus of Messana, with whom he is thought to have studied grammar, music, gymnastics, and perhaps other subjects. His most famous educational connection was with the great Greek philosopher, Socrates, with whom he is thought by some historians to have had a close personal relationship of the type that led to Socrates' conviction of immorality and death. In any case, Plato honored Socrates throughout his life to the point that in one of his letters, he writes that "there is no writing of Plato, nor will there be; the present [writings] are the sayings of a Socrates become beautiful and new."

As a young adult, Plato traveled widely, visiting Egypt, Italy, Sicily, and Cyrene, a Roman colony in modern-day Libya. He appears to have been very much interested in political issues, both in his native Athens and in Syracuse, where he became involved in a political fight that almost resulted in his losing his life. Back in Athens, he founded one of the earliest organized educational institutions known as the Academy, named after a piece of land on which it was founded, which in turn was named after a Greek mythological hero, Akademos (or Academos or Hekademos or Hecademos). Plato died in 348 or 347 BCE under unknown circumstances that have been explained in different ways by different historians.

Plato, like Socrates, was interested throughout his life in developing a philosophy of human life, with special attention to the meaning of the soul and the individual's relationship with the state and the cosmos in general. He worked through this theme in a number of works in his early and middle life, including *The Republic, Phaedo, The Symposium, Phaedrus, Timaeus,* and *Philebus*. His masterpiece was also his longest work, *Laws,* completed just before his death and consisting of 12 volumes. The work takes the form of a discussion among three scholars on a pilgrimage to the terrestrial home of the primary god of the Greeks, Zeus. It deals with issues such as the supernatural source of human laws; the meaning and nature of natural law and natural rights; the relationship of the arts, politics, and religion; and the role of the arts in education.

Project on Scientific Knowledge and Public Policy

School of Public Health & Health Services
The George Washington University
2175 K St., NW Suite 500
Washington, DC 20037
Phone (202) 994–2160
Fax (202) 994–3773
Phone (202) 994–2160
Fax (202) 994–1850
E-mail: sphhsmedia@gwu.edu
URL: http://www.defendingscience.org/

The Project on Scientific Knowledge and Public Policy (SKAPP) was formed in 2001 at the School of Public Health & Health Services (SPHHS) at George Washington University for the purpose of examining the nature of science and studying the way that science is used and misused in government decision-making and legal proceedings. The organization has been sponsored not only by the university, but also by a number of other organizations, including the Open Society Institute, the Public Welfare Foundation, the Rockefeller Family Fund, the Common Benefit Trust (a fund established pursuant to a court order in the Silicone Gel Breast Implant Products Liability litigation), the Alice Hamilton Fund, and the Bauman Foundation. The project staff consists of five individuals, including David Michaels, professor and interim chair of the Department of Environmental and Occupational Health at SPHHS and formerly Assistant Secretary for Environment, Safety and Health at the Department of Energy from 1998 to 2001, and Susan Wood, research professor at SPHHS, formerly deputy director and science advisor for the Congressional Caucus for Women's Issues from 1990 to 1995, director of the Division of Policy and Program Development at the Office of Women's Health of the U.S Department of Health and Human Services (HHS) from 1995 to 2000, and assistant commissioner for Women's Health and Director of the Office of Women's Health at the U.S. FDA from 2000 to 2005. The SKAPP advisory committee consists of a group of eight scholars from a variety of U.S. institutions of higher education, including Boston University, the University of Cincinnati, Tufts University, and the University of Massachusetts at Lowell.

SKAPP's primary work involves the study of specific instances in which government or public policies and/or actions have involved the misrepresentation or misuse of scientific information. To date, the organization has conducted about a dozen such studies on topics such as lung disorders among popcorn workers, bisphenol A, beryllium, diesel particulates, child farm workers, hexavalent chromium, respirator fit testing, perfluorooctanoic acid, and phenylpropanolamine.

The case of the American Conference of Governmental Industrial Hygienists (ACGIH) and its congressional critics is an example of SKAPP's work. Since 1971, ACGIH has been developing and promulgating a group of so-called threshold limit values (TLVs) for workplace exposures to hazardous chemicals. These TLVs have always been voluntary, but have been used by governmental agencies to establish reasonable standards for health risks among American workers.

In about 2000, U.S. Congressman Charlie Norwood (R-Ga.) began a campaign to prevent ACGIH standards from being used by governmental agencies, arguing that setting such standards was the responsibility of governmental agencies, not nongovernmental entities, such as ACGIH. Norwood fought this battle, apparently at least partially on behalf of a number of industries that were concerned that the rigorous ACGIH health standards would encourage workers and the organizations concerned for their health to push for more rigorous health conditions in the workplace. The debate over this issue was finally resolved in 2008 (a year after Representative Norwood had died) when a court ruled that ACGIH had not violated Georgia state law by recommending safety standards for certain types of hazardous work places. Plaintiffs in that case had been International Brominated Solvents Association, Aerosafe Products, Inc., Anchor Glass Container Products, Inc., and the National Mining Association. (Details of this case are available at http://www.defendingscience.org/case-studies/acgih-threshold-limit-values.)

In addition to a review of its major cases, SKAPP provides a useful review of major news stories from 2005 to the present day on its web site along with a page containing some very useful and interesting special features, such as discussions on "Science in the Courts," "Science in Government Decision-Making," "Conversations with Scientists," "Coronado Conference Papers," and "The Pump Handle Blog." The Conversations with Scientists section provides interviews with scientists who present their personal perspectives on issues that arise in the intersection of scientific research and government and politics. The Coronado Conference Papers contains papers presented at the organization's annual meetings on science and politics, while the Pump Handle Blog is an interactive web site that allows participants to exchange news and views on a wide variety of topics, such as life and physical sciences, the environment, politics, medicine, brain and behavior, technology, and information science. Some recent points of discussion have centered on the effect of the government shutdown of October 2013 on women, infant, and children clinics; issues involving transparency in worker safety rulemaking; coal industry refusals to provide assistance to worker widows; effects of austerity measures at the Occupational Safety and Health Administration (OSHA); and health hazards at a Wisconsin iron mine.

Public Employees for Environmental Responsibility

2000 P St., NW, Suite 240
Washington, DC 20036
Phone: (202) 265–PEER (7337)
Fax: (202) 265–4192
E-mail: http://www.peer.org/about-us/contact-us.html
URL: http://www.peer.org/

Public Employees for Environmental Responsibility (PEER) is a 501(c)(3) nonprofit organization of local, state, and federal environmental and natural resource professionals. The organization was founded in 1996 with a four-prong set of objectives in mind: to provide a broad-based support system for professionals in the field of environmental science at all levels; to act as a watchdog for the public interest in issues involving the use of land and other natural resources; to inform and educate the U.S. Congress and other legislative and administrative agencies about important natural resource issues; and to support and advocate for potential whistle-blowers who feel the need to speak out against actions that they deem to be harmful to the nation's natural resources. The organization maintains a national office in Washington, D.C., with a permanent primary staff of five, which is responsible to a board of directors of four permanent members. The organization also acknowledges a small group of about two dozen individuals, whom it refers to as the "Pillars of PEER," who have been involved with the organization for many years and have committed time, resources, and energy to the success of the organization. PEER also maintains regional offices in California, Florida, New England (at North Easton, Massachusetts), New Jersey, the Rocky Mountains (in Denver), the Southwest (in Tucson, Arizona), and Tennessee.

In its two decades of existence, PEER has achieved a number of significant advances on behalf of the nation's national resources and the men and women who work in the management of those resources. Some examples include the following:

- Convinced potential whistle-blowers to allow PEER to take the lead in filing and following up on claims that might otherwise have had deleterious effects on the careers of individual employees

- Successfully defended a number of whistle-blowers in their cases against federal, state, and local agencies, obtaining positive verdicts for their clients' cases, while managing to save their jobs for them

- Developed legislation that was eventually adopted by the U.S. Congress for added protection of National Wildlife Refuges against possible damage by oil and gas drilling projects

- Prevented potential cuts to the nation's tsunami warning system in the FY2013 budget, reducing the risk to individuals living in regions where such storms are likely to occur

- Established the Climate Science Legal Defense Fund, a program designed to protect individual scientists against legal attacks because of their research on global climate change issues

- Defended scientists who had been criticized for their work on issues related to the survival of polar bears in the Arctic region

- Exposed misrepresentations of scientific data by members of the administration of President Barack Obama with regard to the BP Deepwater Horizon accident in the Gulf of Mexico in 2010
- Conducted a vigorous and at least partially successful campaign to prevent the growing of genetically modified crops on National Wildlife Refuge lands

Membership in PEER is open to current and retired public employees, individuals both within and outside government who are sympathetic to the goals of the organizations and to all donors to PEER. Membership brings with it a number of benefits, including legal help at no cost for individuals who are in potential whistle-blower situations; pen pal opportunities, which allow members to exchange opinions and experiences with each other; a host of peer-pressure activities that can be used by individual members or the association as a whole, such as free media training sessions and confidential surveys of employees; and assistance of all kinds to individuals who decide to become whistle-blowers on some specific issue. Some of the assistance provided to members of PEER comes in the form of print and electronic publications. Two regular publications online are PEEReview, a quarterly newsletter of news and events, and PEERmail, a bimonthly newsletter of news, updates, and action alerts. Both are available in print or electronic form.

PEER has also published a number of white papers on a wide variety of topics, such as "Rotten at the Corps—The U.S. Army Corps of Engineers Presiding Over the Death of the Florida Keys," "Horses to Slaughter—Anatomy of a Coverup within the BLM" [Bureau of Land Management], "Land of No Return$—Bankruptcy of the BLM Public Domain Forestry Program," "Bureau of Land Mismanagement—Timber Sale Maladministration," "Never Mind NEPA—No Laws, No Science, No Problem for the BLM," "Savage Salvage—The Timber Feeding Frenzy within BLM, Part Three of a Comprehensive Study of the Forestry Program of the Bureau of Land

Management," "Tortoise on the Half-Shell—Environmental Malpractice at Mojave National Preserve," "War of Attrition—Sabotage of the Endangered Series Act by the U.S. Department of Interior," "Uneven Justice—Environmental Prosecutions in the Clinton Administration and U.S. Attorney Environmental Report Card," "Trading Thin Air—EPA's Plan to Allow Open Market Trading of Air Pollution Credits," "Murky Waters—An Inside Look at EPA's Implementation of the Clean Water Act," and "Swan Dive—Trumpeter Swan Restoration Trumped by Politics."

Much of PEER's work is organized into ongoing campaigns on specific issues involving the relationship of science and politics. Examples of those campaigns deal with efforts to improve protection for and attention to whistle-blowers, programs associated with land use and urban sprawl, dealing with the "malfeasance" of public agencies in dealing with a host of public health issues, improvements in policies for dealing with the uses of public lands, and expanded efforts to protect wildlife. A feature of particular interest on the organization's web site is its Federal Watch/State Watch page, which provides current information on the activities of a large number of governmental agencies, including the Army Corps of Engineers, Bureau of Land Management, Department of Justice, Department of Labor, Federal Communications Commission, Forest Service, Fish and Wildlife Service, Geological Survey, all at the national level, along with data and information about state whistle-blower laws.

Ronald Reagan (1911–2004)

At times, the failure of a politician—or a group of politicians—to make use of new knowledge gained from scientific research simply reflects the fact that that new knowledge may be personally uncomfortable or inconvenient for that (or those) politicians. For someone who has believed and taught throughout his or her life, for example, that marijuana is a "gateway" drug, whose use leads to the use of more dangerous substances, new knowledge that contradicts those beliefs would reasonably be considered to be difficult for a person to adopt and espouse. Examples of the denial or ignoring of "uncomfortable" or "inconvenient" new scientific knowledge are rampant in both American history and world history. One such example might be the outbreak of the AIDS crisis in the United States in the early 1980s.

The first official government report on a disease that became known as acquired immunodeficiency syndrome (AIDS) appeared in the June 5, 1981, issue of the U.S. Centers for Disease Control and Prevention's (CDC) *Morbidity and Mortality Weekly Report* (MMWR). At the time, virtually nothing was known about the disease, but the rate at which it spread throughout certain portions of the American population was striking and frightening. By the end of 1981, 159 cases of AIDS had been reported in the United States, a number that climbed to 771 by the end of 1982, 2,807 in 1983, 7,239 in 1984, and 15,527 in 1985. At that point, 12,529 people with AIDS had died of the disease or related causes. Those numbers continued to climb over the next decade, reaching 160,969 cases and 120,453 deaths in 1990 and 513,486 cases and 319,849 deaths in 1995.

It was apparent early on the epidemic from the lack of knowledge about the disease, its rapid spread, and its high mortality rate that a significant campaign would be necessary to determine the cause of the disease, develop methods of prevent and educational programs, and, if possible, to find a vaccine against the disease. As has traditionally been the case in such instances, those at risk for the disease, health researchers, and other interested individuals called on the federal government to take the lead in the battle against AIDS.

That response was slow to come. For reasons that are not entirely clear, President Ronald Regan, who served in that office from 1981 to 1989 largely ignored the issue of the AIDS crisis. He did not, in fact, mention the word "AIDS" in public until September 17, 1986, more than five years after the crisis began, in a letter to Congress, in which he declared that his administration planned to make AIDS policy a "priority" in coming years. For whatever reason, the Reagan administration essentially ignored the AIDS epidemic for a lengthy period of time, during which, as critics often point out, more Americans died from the disease than died in the Vietnam War. Nor is it possible to say what personal attitudes Reagan may have had about same-sex relationships or AIDS that would have prompted him to ignore the disease for so long. One popular belief is that Reagan himself was uncomfortable with and disapproving of same-sex relationships and that, upon one occasion, he privately made the comment that "[m]aybe the Lord brought down this plague [because] illicit sex is against the Ten Commandments," a belief that was certainly common among many religious and political leaders at the time. (The quote comes from Edmund Morris, *Dutch: A Memoir of Ronald Reagan*. New York: Random House, 1999.)

Any number of factors might explain Reagan's reluctance to deal with AIDS during the majority of his term in office, including personal feelings and the need to placate some portion of the population to which he owed his election and on whose support he depended for success in office. All that remains is the question as to how such a major medical crisis could so long be ignored by a U.S. presidential administration.

Ronald Wilson Reagan was born in Tampico, Illinois, on February 6, 1911, and raised in nearby Dixon. He earned his bachelor of arts degree from Eureka College in Eureka, Illinois, in 1932, and then moved to Davenport, Iowa, where he got a job rebroadcasting baseball games for the Chicago Cubs. He soon developed national recognition for his skill in this specialized field of work. In 1937, Reagan enlisted in the U.S. Army Reserve, and was shortly thereafter discovered by an agent for Warner Brother motion picture studios. He signed a seven-year contract with the company and in 1940 completed the film for which he is perhaps best known, "Knute Rockne—All

American." Reagan went on to appear in nearly 70 other films, most with Warner Brothers, and some in which he was a narrator only. During World War II he made more than 400 training films with the 1st Motion Picture Unit of the U.S. Army Air Force in Culver City, California.

Although he had been a registered Democrat and supported Democratic candidates in most elections, he changed his part affiliation to Republican in 1962 and offered his support to candidate Barrie Goldwater in the 1964 election for the Presidency. He decided to become a candidate for governor himself in 1966, and handily defeated Democratic incumbent Edmund G. ("Pat") Brown for the office. He was elected to a second term in 1970, and then devoted his time to writing and speaking on important national issues, in preparation for a run for the Presidency in 1976. He lost that nomination to Gerald Ford, but earned the party's nomination for the presidency in 1980. He was elected in a landslide over President Jimmy Carter, with 489 electoral college votes to 49 for Carter. He was then reelected in 1984 by an even larger margin over Walter Mondale, who won only Minnesota and the District of Columbia.

After leaving the presidency, Reagan remained active in support of a number of his favorite projects, including the Brady Bill on gun control, the line-item veto, and repeal of the 22nd Amendment to the U.S. Constitution. Reagan was diagnosed with Alzheimer's disease in 1994 and grew progressively worse over the last decade of his life. He died in Los Angeles on June 5, 2004.

Eugenie C. Scott (1945–)

One of the most controversial ongoing issues in American education over the past century has been the teaching of evolution in schools, most often, the teaching of evolution in high school biology classes. On the one hand, there is virtual unanimity among professional biologists of the significance of evolutionary theory to their science. Typical of the positions take on this issue by professional biologists is a 1972 statement, reaffirmed in 1994, by the American Institute of Biological Sciences, which says that "[t]he theory of evolution is the only scientifically defensible explanation for the origin of life and development of species. . . . Biologists may disagree about the details of the history and mechanisms of evolution. Such debate is a normal, healthy, and necessary part of scientific discourse and in no way negates the theory of evolution."

In spite of this view of evolution among professional biologists, many nonbiologists hold other beliefs about the origin of life on Earth and its development (or lack of development) over the ages. These individuals and the groups they have founded have long pushed for the presentation of alternative views of life in biology classes, views known as creationism, scientific creationism, intelligent design, and the like. The battle between the two sides in this argument continues in the second decade of the 21st century as vigorously as it has for nearly a century. The one person most responsible for presenting the scientific view of evolution before state curriculum committees, school groups, and the general public is Eugenie C. Scott, director of NCSE.

Eugenie Carol Scott was born in La Crosse, Wisconsin, on October 24, 1945. She attended the University of Wisconsin at Milwaukee, from which she received her BS and her MS in physical anthropology in 1967 and 1968, respectively. She then enrolled at the University of Missouri at Columbia, from which she earned her PhD in physical anthropology in 1974. Her first teaching assignment was as an instructor in the Extension Division at California State College at Hayward (1968–1970), after which she served as assistant professor of anthropology at Sonoma State College (1970–1971), teaching assistant in anthropology at the University of Missouri (1971–1974), assistant professor of anthropology at the University of Kentucky (1974–1982), and assistant professor of anthropology at the University of Colorado (1984–1986). She was also visiting professor in anthropology at Kansas University in

1976–1977 and postdoctoral scholar in medical anthropology at the University of California at San Francisco in 1983–1984.

In 1986, she was offered the position of director of NCSE in Oakland, California. The position was created with funds provided by the Carnegie Institution and a group of other charitable foundations for the purpose of expanding the work of the organization. Scott retired from her position as executive director of NCSE on January 6, 2014, and became chair of the organization's Advisory Council. In her former position, Scott not only managed the day-to-day activities of the NCSE, but also appeared regularly as a speaker before all types of groups and as a representative for the teaching of evolution on radio and television and before committees of the U.S. Congress and other policy-making groups at the national, state, and local levels.

Scott is the editor of a basic book on the evolution-creationism debate, *Evolution vs. Creationism: An Introduction* (Greenwood Press, 2004) and coeditor of *Not in Our Classrooms: Why Intelligent Design Is Wrong for Our Schools* (Beacon Press, 2006). She is also the author of more than two dozen refereed papers on anthropology and science education, more than 40 chapters, and nonrefereed papers in other journals. She is also the author of nearly 20 books and film reviews on topics in the fields of anthropology and science education. She has been a consultant to a number of organizations on the topic of the teaching of evolution, including the American Civil Liberties Union, Americans United for Separation of Church and State, Southwest Consortium for the Improvement of Mathematics and Science Teaching, Fordham Foundation, San Diego Museum of Man, Templeton Foundation, National Geographic Society, the SETI Institute (Voyages Through Time Project), National Academy of Sciences (Working Group on Teaching Evolution; Steering Committee on Science and Creationism), National Science Foundation, California State Department of Education, California Science Implementation Network, Orange County Department of Education Schools Legal Service, California Academy of Sciences, Science Education for

Public Understanding Program, American Association of Publishers, and the U.S. Civil Liberties Commission.

Among the many honors that Scott has received have been the Defense of Science Award of the Center for Inquiry, Public Service Award of the National Science Board, Distinguished Service Award of the California Science Teachers Association, Public Service Award of the Geological Society of America, Bruce Alberts Award of the American Society for Cell Biology, James Randi Award (Skeptic of the Year) of the Skeptics Society, Hugh H. Hefner First Amendment Award of the Playboy Foundation, and Isaac Asimov Science Award of the American Humanist Association. Scott has also been awarded nine honorary doctorates, from Chapman University, Colorado College, McGill University, Mt. Holyoke College, Ohio State University, Rutgers University, and the universities of Missouri-Columbia, New Mexico, and Wisconsin-Milwaukee. She served as president of the American Association of Physical Anthropologists from 2000 to 2002 and was elected a fellow of the AAAS in 2003.

Kathleen Sebelius (1948–)

One of the most difficult features of any issue involving science and politics is that such issues often do not fall neatly into one that requires a strictly scientific decision or one that involves strictly a political decision. Many issues involve both technical (scientific) elements and policy (political) elements. How should such decisions be made?

An example of such a case was the ongoing controversy over the so-called morning after contraceptive drug known as Plan B throughout most of the first decade of the 21st century. The safety and efficacy of the drug was studied in great deal by scientists, and government committees responsible for deciding as to whether it should be offered to all women without prescription almost unanimously favored such an action. But for a number of years, government officials with the final say on such issues decided to ignore this scientific advice and, largely for political reasons, decided not to grant such approval. The person who was ultimately responsible for these decisions was Secretary of HHS, Kathleen Sebelius. In 2011, for example, Sebelius overruled a decision by the FDA and Center for Drug Evaluation and Research (CDER) saying that data provided by the drug manufacturer "do not conclusively establish that Plan B One-Step should be made available over the counter for all girls of reproductive age." Sebelius did not say that the drug was unsafe or not efficacious, just that there were not enough data to permit it to be sold on a nonprescription basis to women under the age of 17. Two years later, a federal judge disagreed with the Secretary and ordered the FDA to approve OTC sale of Plan B to women of all ages. This controversy is all the more interesting because in almost all other cases, Sebelius had been seen as a strong advocate for women's causes, such as contraceptive use, and her decisions in this case made her a whole new set of political foes over a scientific issue.

Kathleen Sebelius (née Gilligan) was born in Cincinnati, Ohio, on May 15, 1948, to Mary Kathryn Dixon Gilligan and John Joyce Gilligan. John Gilligan, better known as "Jack," served as U.S. Representative from the first district of Ohio from 1965 to 1967 and as governor of the state from 1971 to 1975. Kathleen attended the Summit Country Day School in Cincinnati before matriculating at Trinity Washington University in Washington, D.C., from which she graduated with a bachelor of arts degree in political science in 1970. She moved to Kansas in 1974 and continued her studies at the University of Kansas, which granted her a master of business administration (MBA) degree in 1977.

After completing her MBA studies, Sebelius took a position as executive director and chief lobbyist for the Kansas Trial Lawyers Association (now the Kansas Association for Justice), a position in which she served until 1986. She then ran for and was elected to a seat in the Kansas House of Representatives, where she was reelected three more times. In 1994 she left the

House to run for state Insurance Commissioner, a campaign that she also won. In 2002, Sebelius ran for and was elected to the office of governor of the State of Kansas, the first occasion in American history in which a father and daughter pair had both served as governors of U.S. states. Sebelius was reelected in 2006, and served in that post until she was nominated to be secretary of HHS by newly elected president Barack Obama in March 2009. She continues to hold that post as of 2014.

B. F. Skinner (1904–1990)

One of the great debates in the history of science is what responsibility, if any, scientists have for the use to which society puts their discoveries. One of the men constantly torn by that question was Alfred Nobel, the discoverer of dynamite. Nobel well understood the enormous benefits that dynamite brought to his fellow humans in fields such as mining and construction. But he was also well aware of the potentially devastating effects produced by the use of dynamite in warfare. He tried to resolve this issue by endowing the Nobel Prizes, which recognize signal achievements in physics, chemistry, biology, literature, and peace. Other scientists have found other ways to ensure that society makes the best possible use of their discoveries or, from an entirely different philosophical perspective, decided that such a decision was not theirs to make; their job was to discover, society's job was to decide how best to use those discoveries.

Psychologist B. F. Skinner was never in a position to make wholesale societal changes that reflected his discoveries. However, he had very definite ideas as to how his work could change and improve human life on a number of levels, from the raising of children and the teaching of students at all levels, to the grand restructuring of human society to bring the greatest good to the greatest number of citizens.

Skinner's field of interest was learning. How do organisms learn to behave in one way when confronted with a particular situation, rather than in some very different way? Many people believe that a major part of the answer to that question is free will. People (and perhaps other animals) *choose* to do A or B when confronted with situation X. Skinner rejected that notion. He believed that learning was entirely a stimulus—response—reward event. When an organism is presented with some type of situation, such as the choice between opening one of two doors with unknown objects behind them, the organism will choose one or the other on a somewhat random basis. The reward or punishment the organism receives after it makes that choice determines what its behavior is likely to be in the future. For example, a pigeon that pecks at a red-colored door and receives a grain of corn is more likely to peck at a red-colored door in the future.

Skinner carried out very large numbers of experiments, most famously with pigeons, but also with other animals, to find out what the rules of learning were for this type of behavior. He discovered, for example, that an animal learns better if it is rewarded on an irregular schedule, rather than being rewarded every time or on a regular pattern.

Skinner saw all number of applications of his discoveries. One of the most famous was a so-called *Skinner box*, a device into which an organism could be put and taught by rewarding a set of responses that an experimenter determined to be "good." Skinner used such a box to help train his own daughter. Although criticized for being inhumane, the Skinner box had many positive advantages and greatly simplified and improved methods for the raising of children.

To some extent, Skinner is most famous today for his novel *Walden Two*, a book in which he describes a utopian society that operates on the principles of Skinnerian psychology. Specifically, each individual in society is chose at birth to play some particular role in the culture—teacher, political office holder, gardener, nurse, or other—and then trained by a response and reward system to learn that role and to love playing it. Such a society, Skinner believed, would make the greatest number of people happy in a well-functioning society. In the two decades after the publication of *Walden Two*, at least nine efforts were made to create real communities based on Skinner's plan. Of all these efforts, only one, Los Horcones, in New Mexico, continues to function as a Walden Two–type community.

Burrhus Frederic (B. F.) Skinner was born on March 20, 1904, in Susquehanna, Pennsylvania. He attended Hamilton College, in Clinton, New York, where he majored in English literature, with plans of becoming a writer. He did not fit well into the structured environment at Hamilton, but did graduate in 1926 with his bachelor of arts degree. He then took a year off to live at home at try his hand at writing "the great American novel," at which he was a complete failure. He struggled for the next few years, not knowing exactly what to do with his life, until he came across the works of some early leaders in the new science of psychology, including John Watson and Ivan Pavlov. He learned enough of the subject to decide to continue his studies at Harvard in the field of psychology. He earned his master's degree there in 1930 and his PhD a year later. He then stayed on at Harvard as a research assistant.

In 1936, Skinner accepted an appointment in the psychology department at the University of Minnesota, where he remained for nearly a decade. In 1945, he moved to the University of Indiana, where he served as chair of the Department of Psychology from 1946 to 1947. A year later, he resigned his post at Indiana to return to Harvard, where he remained for the rest of his academic career. Skinner died at his home in Cambridge, Massachusetts, on August 18, 1990. He is widely regarded as one of the most famous names in the history of psychology, often being compared to Sigmund Freud himself. Among his many honors are the Howard Crosby Warren Medal of the Society of Experimental Psychologists, Edward Lee Thorndike Award of the American Psychological Association, National Medal of Science of the National Science Foundation, Gold Medal Award of the American Psychological Foundation, Humanist of the Year Award of the American Humanist Society, William James Fellow Award of the American Psychological

Society, and Lifetime Achievement Award of the American Psychology Association. Skinner also received 20 honorary doctorates from institutions ranging from Alfred University and Hamilton College to the University of Chicago and Harvard University.

Union of Concerned Scientists

Two Brattle Square
Cambridge, MA 02138–3780
Phone: (617) 547–5552
Fax: (617) 864–9405
E-mail: http://www.ucsusa.org/about/public-information-request.html
URL: http://www.ucsusa.org/

The UCS is an organization of scientists, engineers, and lay persons working together to develop technical solutions for world problems and to educate politicians, legislators, and the general public about the nature of those problems and some of the ways in which those problems can be addressed. The organization was founded in 1969 as a collaboration between students and faculty at the Massachusetts Institute of Technology (MIT) at a time when public awareness of important socioscientific issues had just begun to grow throughout the United States and many other parts of the world. One initial impetus in the creation of the group was concern about the ways in which scientific and technological knowledge was being used for the creation of weapons of destruction in the Vietnam War. Another important factor was the growing awareness among the American public of the profound and extensive ecological and environmental issues that had developed in the United States, in many cases, largely as a consequence of scientific and technological developments in preceding decades.

Some of the early programs in which the UCS became involved were critiques of the Nixon administration's plans for the design and construction of an antiballistic missile system, providing information to the general public about the risks of nuclear power generation following the disaster at the Three Mile Island nuclear power plant in 1979, and becoming advocates in opposition to President Ronald Reagan's plans for a Strategic Defense Initiative, commonly known as the Star Wars defense program.

Today, UCS claims a membership of more than 400,000 scientists and nonscientists. Policy is set by an 18-member board of directors that always includes some of the most famous names in science, such as, in recent years, Richard L. Garwin, winner of the National Medal of Science; Nobel Prize winner Mario Molina; James J. McCarthy, former director of the Museum of Comparative Zoology at Harvard University; Peter A. Bradford, former member of the Nuclear Regulatory Commission; and Ellyn R. Weiss, artist, curator, and former assistant attorney general for environmental affairs for the Commonwealth of Massachusetts. In addition to its national headquarters in Cambridge, UCS has regional offices in Washington, D.C.; Berkeley, California; and Chicago. The organization has a very large technical staff organized into special areas of expertise, including the Center for Science and Democracy, clean energy, clean vehicles, communications, development, global warming, nuclear power, nuclear weapons and global security, food and agriculture, and scientific integrity.

The Center for Science and Democracy was established in 2012 as the result of a $1 million gift from American physicist and former head of the National Bureau of Standards, Lewis M. Branscomb. The center was established because of growing concerns about the way in which the general public as well as legislators and other politicians view science and make use of (or ignore) the scientific data that are available for dealing with a host of socioscientific problems in everyday life. One objective of the new center is to bring together experts, decision makers, and the public together to discuss and debate key issues "at the intersection of science and democracy, including special interests' influence on science used in government decision making."

Predating the creation of the Center, the UCS established its Scientific Integrity program in 2004 in response to concerns about the problem of interference by government in the scientific enterprise. Over the ensuing decade, the organization focused on efforts to strengthen scientific integrity in the federal government, work to protect and expand government safeguards on air, water, food, biodiversity, and other science-related issues, and act to protect scientists whose work and personal status was under attack by government. Out of this work have come a number of important reports, such as *Grading Government Transparency*, *Federal Agency Scientific Integrity Policies*, *Science in the Age of Scrutiny*, and *A Climate of Corporate Control*.

Three key publications available from UCS are its magazine, *Catalyst*, which is published three times a year and carries articles on important socioscientific topics, current news, a perspectives column, an interview, and a member profile; *Earthwise*, a quarterly newsletter that contains regular features such as Fast Facts, Take Action items, UCS on the Web, and updates on current topics of interest; and *Got Science?*, an electronic newsletter that contains discussions of important socioscientific issues in the news. In addition, the organization has published a large number of books, booklets, brochures, reports, and other items on issues in which the UCS is interested and involved. Some examples are *Cooler Smarter: Practical Steps for Low-Carbon Living*; *Wood for Good: Solutions for Deforestation-Free Wood Products*; *Grade A Choice? Solutions for Deforestation-Free Meat*; *Recipes for Success: Solutions for Deforestation-Free Vegetable Oils*; *Heat in the Heartland: 60 Years of Warming in the Midwest*; *After the Storm: The Hidden Health Risks of Flooding in a Warming World*; *Rising Temperatures, Worsening Ozone Pollution*; *Logging and the Law: How the U.S. Lacey Act Helps Reduce Illegal Logging in the Tropics*; and *Smoke,*

Mirrors and Hot Air: How ExxonMobil Uses Big Tobacco's Tactics to Manufacture Uncertainty on Climate Science. A very useful source of information and opinions on the interaction of science, politics, and government is the UCS Blog, on its web site at http://blog.ucsusa.org/.

The Palisades are NOT For Sale
LG Tower
A Monument to Corporate Arrogance

Introduction

The interaction of science and politics has been the subject of a great deal of thought, legislative action, court cases, and other commentary over the centuries. This chapter contains extracts from some of the most important of these documents. It also provides some basic data related to the interaction of science and politics in the United States and around the world.

Data

Table 5.1. In 2007, the research group Mathematica Policy Research, Inc., conducted a study for the U.S. Department of Health and Human Services on the effectiveness of four abstinence-only education programs comparing participants in those programs with a matched control group. Among the measures for which the two groups were tested are those in the tables shown here.

Protestors rally at the groundbreaking of the new LG Electronics headquarters in Englewood Cliffs, New Jersey, in November, 2013. The group announced an intensified campaign to stop LG from raising the first building ever to go above the tree line in the Palisades. (Jonathan Fickies/ AP Images for Protect the Palisades)

Table 5.1   Impact of Abstinence-Only Education Programs on Behavior

Number of Sex Partners (Percentage: Program group/control group)

Measure	Four Programs	MCMF	RCV	FUPTP	TIC
Always Abstinent	49/49	38/38	44/40	60/62	53/57
One Partner	16/16	21/15	19/24	12/10	13/15
Two Partners	11/11	9/15	13/13	10/10	10/7
Three Partners	8/8	8/11	10/11	5/3	7/7
Four or More Partners	17/16	22/21	13/13	14/16	17/15

Four programs = Total of all four programs
MCMF = My Choice, My Future!
RCV = ReCapturing the Vision
FUPTP = Families United to Prevent Teen Pregnancy
TIC = Teens in Control

Condom Use; First Sexual Experience (Percentage: Program group/control group)

Measure	Four Programs	MCMF	RCV	FUPTP	TIC
Always Abstinent	49/49	38/38	44/40	60/62	53/57
Condom Use	44/43	53/53	52/53	36/32	36/35
No Condom Use	7/8	9/9	4/7	5/6	10/8

Condom Use; Last 12 Months (Percentage: Program group/control group)

Measure	Four Programs	MCMF	RCV	FUPTP	TIC
Always Abstinent	49/49	38/38	44/40	60/62	53/57
Always	23/23	25/25	48/43	65/67	66/68
Sometimes	17/17	24/24	21/23	12/11	12/11
Never	4/4	7/7	7/5	2/2	3/2

Use of Birth Control, First Sexual Experience (Percentage: Program group/control group)

Measure	Four Programs	MCMF	RCV	FUPTP	TIC
Always Abstinent	49/49	38/38	44/40	60/62	53/57
Yes	45/44	56/54	53/54	36/32	37/36
No	6/7	6/8	3/7	4/6	10/7

Use of Birth Control, Last 12 (Percentage: Program group/control group)

Measure	Four Programs	MCMF	RCV	FUPTP	TIC
Always Abstinent	49/49	38/38	44/40	60/62	53/57
Always	29/29	40/40	31/33	25/23	21/20
Sometimes	13/14	14/14	18/20	9/10	11/10
Never	2/2	1/3	3/3	1/1	2/2

Perceived Effectiveness of Condoms for Preventing Pregnancy (Percentage: Program group/control group)

Measure	Four Programs	MCMF	RCV	FUPTP	TIC
Usually	51/52	56/60	57/57	47/47	44/44
Sometimes	38/38	41/39	34/34	37/38	41/40
Never	3/3	1/0	4/3	4/3	4/4
Unsure	7/7	2/1	4/6	12/12	11/11

Perceived Effectiveness of Condoms for Preventing HIV (Percentage: Program group/control group)

Measure	Four Programs	MCMF	RCV	FUPTP	TIC
Usually	34/38	27/41	44/47	38/28	28/37
Sometimes	30/30	37/31	30/28	22/33	31/29
Never	21/17	31/18	15/16	20/16	18/16
Unsure	14/15	5/11	10/9	20/22	23/19

Source: "Impacts of Four Title V, Section 510 Abstinence Education Programs." Princeton, NJ: Mathematica Policy Research, April 2007, Christopher Trenholm, Barbara Devaney, Ken Fortson, Lisa Quay, Justin Wheeler, and Melissa Clark. Data reproduced by permission.

Table 5.2. During the first decade of the 21st century, the Union of Concerned Scientists (UCS) conducted a number of surveys of scientists working in and for the federal government on the topic of the politicization of science. In 2011, the UCS conducted a follow-up survey of scientists working at the Food and Drug Administration (FDA) similar to an earlier survey conducted in 2006. Some of the most important findings of the 2011 are reported here.

Table 5.2   **Survey of FDA Scientists. Percent of Scientists Expressing Each Opinion on Each Question**

Opinion	Q7*	Q8	Q9	Q11	Q12	Q13	Q14	Q15
Strongly Agree	8.9	26.6	12.3	13.4	7.5	2.7	29.1	7.6
Agree	29.5	34.6	36.7	40.3	22.9	12.6	47.1	29.8
Don't Know	38.1	19.9	18.9	20.0	47.9	43.0	12.0	35.4
Disagree	17.3	14.0	22.1	19.2	14.9	29.2	10.0	22.5
Strongly Disagree	6.3	4.9	9.9	7.1	6.7	12.5	1.8	4.8

(Continued)

Table 5.2 *(Continued)*

Opinion	23a	23b	23c	23d	23e	23f	23g
Frequently	7.6	5.3	3.6	7.3	3.1	4.6	3.2
Occasionally	29.5	23.5	20.1	23.4	12.1	15.4	10.9
Seldom	17.5	16.5	21.0	17.6	14.4	20.7	16.6
Never	21.5	28.5	27.8	22.9	44.6	37.1	43.8
Not Applicable	23.9	26.1	27.4	28.8	25.8	22.2	25.5

*Question numbers correspond to questions asked in UCS survey. See the list of questions below.

Survey questions:

7. FDA leadership stands behind agency employees or managers who make decisions that may be controversial.

8. My direct supervisor stands behind scientists who put forth positions that may be controversial.

9. The FDA offers opportunity for advancement based on scientific expertise, not just on administrative and supervisory expertise.

11. Currently, I can openly express any concerns about the mission-driven work of my agency without fear of retaliation.

12. Currently, I am allowed to publish work in peer-reviewed scientific journals regardless of the level of controversy of the topic.

13. Currently, I am allowed to speak to the public and the news media about my scientific research findings, regardless of the level of controversy of the topic.

14. FDA leadership is as committed to product safety as it is to bringing products to the market.

15. The current level of independence and authority of FDA postmarket product safety systems sufficiently serves the public.

23a. Public health has been harmed by businesses withholding safety information from the agency.

23b. Corporate interests have forced the withdrawal or significant modification of a FDA policy or action designed to protect consumers or public health.

23c. Nongovernmental interests (such as advocacy groups) have forced the withdrawal or significant modification of a FDA policy or action designed to protect consumers or public health.

23d. Members of Congress have forced the withdrawal or significant modification of a FDA policy or action designed to protect consumers or public health.

23e. FDA decision makers made requests to inappropriately exclude or alter technical information or conclusions in a FDA scientific document.

23f. Selective or incomplete use of data was used to justify a specific regulatory outcome.

23g. Changes or edits were made during review that altered the meaning of scientific findings without providing a meaningful opportunity to correct them.

Source: "2011 Survey of FDA Scientists: Detailed Response Data." Union of Concerned Scientists. http://www.ucsusa.org/assets/documents/scientific_ integrity/fda-2011-detailed-survey-data.pdf. Accessed on July 31, 2013. Data used with permission of the Union of Concerned Scientists.

http://www.ucsusa.org/assets/documents/scientific_integ rity/fda-2011-detailed-survey-data.pdf

Documents

The Republic, by Plato (ca. 389 BCE)

In his famous work, The Republic, the Greek scholar Plato lays out the justification under which leaders of a nation may lie to their citizens: when the greater good of the society allows and even demands such actions. His explanation follows. (Ellipses, . . ., designate omitted passages.)

Then if any one at all is to have the privilege of lying, the rulers of the State should be the persons; and they, in their dealings either with enemies or with their own citizens, may be allowed to lie for the public good. . . .

How then may we devise one of those needful falsehoods of which we lately spoke—just one royal lie which may deceive the rulers, if that be possible, and at any rate the rest of the city?

What sort of lie? he said.

Nothing new, I replied; only an old Phoenician tale (Laws) of what has often occurred before now in other places, (as the poets say, and have made the world believe,) though not in our time, and I do not know whether such an event could ever happen again, or could now even be made probable, if it did. . . .

Well then, I will speak, although I really know not how to look you in the face, or in what words to utter the audacious fiction, which I propose to communicate gradually, first to the rulers, then to the soldiers, and lastly to the people. They are to be told that their youth was a dream, and the education and training which they received from us, an appearance only; in reality during all that time they were being formed and fed in the womb of the earth, where they themselves and their arms and appurtenances were manufactured; when they were completed, the earth, their mother, sent them up; and so, their country being their mother and also their nurse, they are bound to advise for her good, and to defend her against attacks, and her citizens they are to regard as children of the earth and their own brothers.

You had good reason, he said, to be ashamed of the lie which you were going to tell.

True, I replied, but there is more coming; I have only told you half. Citizens, we shall say to them in our tale, you are brothers, yet God has framed you differently. Some of you have the power of command, and in the composition of these he has mingled gold, wherefore also they have the greatest honour; others he has made of silver, to be auxiliaries; others again who are to be husbandmen and craftsmen he has composed of brass and iron; and the species will generally be preserved in the children. But as all are of the same original stock, a golden parent will sometimes have a silver son, or a silver parent a golden son. And God proclaims as a first principle to the rulers, and above all else, that there is nothing which they should so anxiously guard, or of which they are to be such good guardians, as of the purity of the race. They should observe what elements mingle in their offspring; for if the son of a golden or silver parent has an admixture of brass and iron, then nature orders a transposition of ranks, and the eye of the ruler must not be pitiful towards the child because he has to descend in the scale and become a husbandman or artisan, just as there may be sons of artisans who having an admixture of gold or silver in them are raised to honour, and become guardians or auxiliaries. For an oracle says that when a man of brass or iron guards the State, it will be destroyed. Such is the tale; is there any possibility of making our citizens believe in it?

Not in the present generation, he replied; there is no way of accomplishing this; but their sons may be made to believe in the tale, and their sons' sons, and posterity after them.

Source: Plato. The Republic. Project Gutenberg EBook. http://www.gutenberg.org/files/1497/1497-h/1497-h .htm#link2H_4_0006. Accessed on July 24, 2013.

———

Whistleblower Protection Act of 1989

Individuals who work for government agencies face a difficult decision when they discover instances in which the agency for which they work operates in an unprofessional, unethical, immoral, or illegal way. An example might be an employee of the Environmental Protection Agency who is asked or told to change the content of her or his report in such a way as to include questionable or incorrect information. If the individual's supervisor makes such a demand, the employee may feel that she or he has few options other than to follow those directions, unless she or he is willing to lose her or his job. One option an employee has is to report the case to an outside authority, such as the Department of Justice. But in doing so, the employee faces the risk of being fired for taking such an action. The U.S. Congress passed the Whistleblower Act of 1989 in an effort to provide employees protection from such retaliatory actions by government agencies. The portion of the act explaining its purpose follows (omissions are designated by ellipses, . . .). This section should be read, however, in conjunction with a relevant Supreme Court ruling on whistle-blowing in 2006.

(a) FINDINGS—The Congress finds that—

   (1) Federal employees who make disclosures described in section 2302(b)(8) of title 5, United States Code, serve the public interest by assisting in the elimination of fraud, waste, abuse, and unnecessary Government expenditures;

   (2) protecting employees who disclose Government illegality, waste, and corruption is a major step toward a more effective civil service; and

   (3) in passing the Civil Service Reform Act of 1978, Congress established the Office of Special Counsel to protect whistle-blowers (those individuals who make disclosures described in such section 2302(b)(8)) from reprisal.

(b) PURPOSE—The purpose of this Act is to strengthen and improve protection for the rights of Federal employees, to prevent reprisals, and to help eliminate wrongdoing within the Government by—

(1) mandating that employees should not suffer adverse consequences as a result of prohibited personnel practices; and

(2) establishing—

(A) that the primary role of the Office of Special Counsel is to protect employees, especially whistleblowers, from prohibited personnel practices;

(B) that the Office of Special Counsel shall act in the interests of employees who seek assistance from the Office of Special Counsel; and

(C) that while disciplining those who commit prohibited personnel practices may be used as a means by which to help accomplish that goal, the protection of individuals who are the subject of prohibited personnel practices remains the paramount consideration.

Source: S.20 Whistleblower Protection Act of 1989. http:// thomas.loc.gov/cgi-bin/query/C?c101:./temp/~c1012XCxNR. Accessed on July 29, 2013

Abortion and Breast Cancer Legislation (1990s)

The question as to whether a woman's risk for breast cancer increases as the result of having had an abortion has been studied by scientists for decades. At this point in history, there is virtually unanimous agreement within the scientific community that no such link exists. Nonetheless, individuals and groups who oppose abortion continue to use this purported link as a reason for enacting laws limiting or prohibiting abortions. Such laws have been proposed and/or passed in a number of states since the 1990s. In 2013, the State of Kansas adopted new abortion legislation that included reference to the supposed link between abortion and breast cancer, as follows (omitted material is indicated by ellipses, . . .):

HOUSE BILL No. 2253

. . .

Sec. 14. K.S.A. 2012 Supp. 65-6709 is hereby amended to read as follows: 65-6709. No abortion shall be performed or induced without the voluntary and informed consent of the woman upon whom the abortion is to be performed or induced. Except in the case of a medical emergency, consent to an abortion is voluntary and informed only if:

(a) At least 24 hours before the abortion, the physician who is to perform the abortion or the referring physician has informed the woman in writing of: . . .

[A number of specific pieces of information are required to be provided by the doctor, one of which is:]

(3) a description of risks related to the proposed abortion method, including risk of premature birth in future pregnancies, risk of breast cancer and risks to the woman's reproductive health, and alternatives to the abortion that a reasonable patient would consider material to the decision of whether or not to undergo the abortion;

. . .

Sec. 15. K.S.A. 2012 Supp. 65–6710 is hereby amended to read as follows: 65–6710. (a) The department shall cause to be published and distributed widely, within 30 days after the effective date of this act, and shall update on an annual basis, the following easily comprehensible informational materials:

. . .

[A list of information to be included in the printed materials is listed here, including:]

The material shall also contain objective information describing the methods of abortion procedures commonly employed, the medical risks commonly associated with each such procedure, *including risk of premature birth in future pregnancies, risk of breast cancer, risks to the woman's reproductive health* and the medical risks associated with carrying an unborn child to term. *[Emphasis in original.]*

Source: House Bill No. 2253. State of Kansas. http://www.kslegislature.org/li/b2013_14/measures/documents/hb2253_enrolled.pdf. Accessed on July 25, 2013.

Politics and Science in the Federal Government (2003)

In 2003, Representative Henry Waxman (R-Calif.) asked the Special Investigations Division of the U.S. House of Representatives Committee on Government Reform to conduct a study of the way in which the Bush administration was dealing with scientists and scientific issues working in and for the federal government. The investigators found that the administration was interfering with the normal flow of scientific information in at least 21 areas, abstinence-only education, agricultural pollution, Arctic National Wildlife Refuge, breast cancer, condoms, international negotiations, drinking water, education policy, environmental health, global warming, HIV/AIDS, lead poisoning, missile defense, oil and gas, prescription drug advertising, reproductive health, stem cells, substance abuse, wetlands, workplace safety, and Yellowstone National Park. They suggested that this interference took a number of forms, including the following.

Manipulating Scientific Advisory Committees

Scientific advisory committees assure that the government hears from the nation's top experts in a particular field before creating policy in that area. The Federal Advisory Committee Act requires that such committees be "fairly balanced in terms of the points of view represented" and requires that advice and recommendations "not be inappropriately influenced by the appointing authority or by any special interest." The Bush Administration, however, has repeatedly manipulated the advisory committee process to advance its political and ideological agenda. Examples include the following:

- Appointing Unqualified Persons with Industry Ties. After dropping three national experts in lead poisoning from the Advisory Committee on Childhood Lead Poisoning Prevention, the Department of Health and Human Services appointed several individuals with ties to the lead industry, including a lead industry consultant who had testified that a lead level seven times the current limit is safe for children's brains.

- Appointing Unqualified Persons with Ideological Agendas. The Department of Health and Human Services nominated as chair of the FDA's Reproductive Health Drug Advisory Committee an antiabortion activist who recommends that women read the bible for relief of premenstrual symptoms. The appointee's principal credential appears to be his opposition to the abortifacient RU-486. The medical journal *Lancet* described his scientific record as "sparse" and wrote that "[a]ny further right-wing incursions on expert panels' membership will cause a terminal decline in public trust in the advice of scientists."

- Stacking Advisory Committees. The Department of Health and Human Services replaced 15 of 18 members of the key advisory committee to the National Center on Environmental Health. Several of the new members were long-time industry consultants. In response, 10 leading scientists wrote in *Science* that "stacking these public committees out of fear that they may offer advice that conflicts with administration policies devalues the entire federal advisory committee structure and the work of dedicated scientists who are willing to participate in these efforts."

- Opposing Qualified Experts. The Department of Health and Human Services rejected a widely respected expert's nomination to a grant review panel on workplace safety after it became clear that she supported rules to protect workers from musculoskeletal injuries, rules that the Bush Administration opposes. The head of the panel called the rejection "directly opposed to the philosophy of peer review, which is supposed to be nonpolitical and transparent."

Distorting and Suppressing Scientific Information

The public relies on government agencies for accurate scientific information, evidence-based decision making on matters of life and health, and clear explanations of complex technical matters. Under the Bush administration, however, administration officials have withheld or skewed important scientific information that conflicts with the Bush administration's ideological and political agenda. Examples include the following:

- Including Misleading Information in Presidential Communications. After banning research on new lines of embryonic stem cells, President Bush assured the American people that research on "more than 60" existing lines cells "could lead to breakthrough therapies and cures." In fact, only 11 cell lines are now available for research, all of which were grown with mouse cells, rendering them inappropriate for treating people.

- Presenting Incomplete and Inaccurate Information to Congress. When Interior Secretary Gale Norton assured Congress that drilling in the Alaska National Wildlife Refuge would not harm the region's caribou population, she altered or omitted multiple key scientific conclusions prepared by federal biologists at the refuge. One Fish and Wildlife Service official commented, "We tried to present all the facts, but she only passed along the ones she liked. And to pass along facts that are false, well, that's obviously inappropriate."

- Altering Web Sites. As social conservatives campaigned to require women to be "counseled" about an alleged risk of breast cancer from abortions, the National Cancer Institute revised its web site to suggest that studies of equal weight conflicted on the question. In fact, there is scientific consensus that no such link exists; as the head of epidemiology research at the American Cancer Society had concluded previously, "This issue has been resolved scientifically. . . . This is essentially a political debate."

- Suppressing Agency Reports. After the White House edited a discussion of global warming in the Environmental Protection Agency's Draft Report on the Environment, agency scientists objected that the draft "no longer accurately represents scientific consensus on climate change," and EPA chose to eliminate the discussion entirely. A former EPA administrator in the Nixon and Ford administrations commented, "I can state categorically that there was never such White House intrusion into the business of the E.PA. *[sic]* during my tenure."

Interfering with Scientific Research

The federal government invests $100 billion annually in scientific research to discover new cures, protect the environment, defend the country, and support other effective policies for the health and welfare of the American people. But instead of encouraging the development and dissemination of objective scientific information, the Bush administration has repeatedly interfered with scientific research and analysis where political and ideological interests are at stake. Examples include the following:

- Scrutinizing Ongoing Research. Officials of the National Institutes of Health warned HIV researchers to expect increased scrutiny of any research grant requests using the words "gay" or "men who sleep with men." The administration has also instituted a new policy at the Agriculture Department requiring scientists to seek approval of any research on "agricultural practices with negative health and environmental consequences."

- Obstructing Agency Analyses. The Bush administration refused to let the Environmental Protection Agency conduct analyses on air quality proposals that differ from the president's "Clear Skies" initiative. William Ruckelshaus, the first EPA administrator under President Nixon, said of this pattern, "Is the analysis flawed? That is a legitimate reason for not releasing it. But if you don't like the outcome that might result from the analysis, that is not a legitimate reason."

- Undermining Outcome Assessment. The Centers for Disease Control and Prevention used to evaluate sex education programs and identify those with scientific evidence of effectiveness. After social conservatives complained that none of the programs taught "abstinence only," the agency ended the "Programs That Work" initiative altogether.

- Blocking Scientific Publication. The Agriculture Department prohibited one of its microbiologists from publishing or presenting research indicating that industrial hog farming may contribute to antibiotic resistance. The scientist traced the department's actions back to communications from industry.

This report describes these and other examples of interference in the scientific process.

> **Source:** United States House of Representatives. Committee on Government Reform—Minority Staff. Special Investigations Division. *Politics and Science in the Bush Administration.* http://oversight-archive.waxman.house.gov/documents/20080130103545.pdf. Accessed on July 25, 2013.

Gil Garcetti, et al., Petitioners v. Richard Ceballos, 547 U.S. 410 (2006)

The United States government has a long history of recognizing the possible need for whistle-blowers and the risks that such individuals might face in the workplace. In 1989, the U.S. Congress passed a Whistleblower Protection Act designed to update and confirm federal protection for government employees who speak out against improper actions taken by the agencies for which they work. The first real test of that law came in the early 2000s when Richard Ceballos, a deputy attorney general in the Los Angeles County District Attorney's office, raised objections about evidence in a case to which he had been assigned. Ceballos later claimed that he was subjected to a number of retaliatory actions by his supervisors because of his actions in that case, and he sued the Los Angeles County District Attorney for the retaliatory actions. The case eventually reached the U.S. Supreme Court, which ruled in a 5 to 4 vote that Ceballos was not protected against retaliation by the 1989 act because he was a public employee. The court majority's reasoning was as follows (omitted text is indicated with ellipses, . . .):

The controlling factor in Ceballos' case is that his expressions were made pursuant to his duties as a calendar deputy. . . . That consideration—the fact that Ceballos spoke as a prosecutor fulfilling a responsibility to advise his supervisor about how best to proceed with a pending case—distinguishes Ceballos' case from those in which the First Amendment provides protection against discipline. We hold that when public employees make statements pursuant to their official duties, the employees are not speaking as citizens for First Amendment purposes, and the Constitution does not insulate their communications from employer discipline.

Ceballos wrote his disposition memo because that is part of what he, as a calendar deputy, was employed to do. It is immaterial whether he experienced some personal gratification from writing the memo; his First Amendment rights do not depend on his job satisfaction. The significant point is that the memo was written pursuant to Ceballos' official duties. Restricting speech that owes its existence to a public employee's professional responsibilities does not infringe any liberties the employee might have enjoyed as a private citizen. It simply reflects the exercise of employer control over what the employer itself has commissioned or created.

A contrary interpretation of the facts in the case was presented in a dissent written by Justice David Souter, who said:

[A government employee] speaking as a citizen, that is, with a citizen's interest, is protected from reprisal unless the statements are too damaging to the government's capacity to conduct public business to be justified by any individual or public benefit thought to flow from the statements. . . . Entitlement to protection is thus not absolute.

This significant, albeit qualified, protection of public employees who irritate the government is understood to flow from the First Amendment, in part, because a government paycheck does nothing to eliminate the value to an individual of speaking on public matters, and there is no good reason for categorically discounting a speaker's interest in commenting on a matter of public concern just because the government employs him. Still, the First Amendment safeguard rests on something more, being the value to the public of receiving the opinions and information that a public employee may disclose. "Government employees are often in the best position to know what ails the agencies for which they work."

[As might be expected, the Court's decision was strongly criticized by a number of organizations who depend on whistle-blowing for much of their information. For example, the legal director of Climate Science Watch, an organization that promotes integrity in climate science research in government, said that:]

This decision is outrageous. Canceling the doctrine of 'duty speech' means that government employees only have an on-the-job right to be 'yes people,' parroting false information and enabling illegality.

Sources: *Garcetti v. Ceballos,* 547 U.S. 410. http://www .supremecourt.gov/opinions/05pdf/04–473.pdf. Accessed on July 29, 2013; Rick Piltz. "Supreme Court Rules against Government Whistleblowers." Climate Science Watch. http://

www.climatesciencewatch.org/2006/05/29/supreme-court-rules-against-government-whistleblowers/. Accessed on July 29, 2013.

_____

Politicization of the Surgeon General's Office (2007)

On July 10, 2007, the House of Representatives Committee on Oversight and Government Reform held a hearing entitled "The Surgeon General's Vital Mission: Challenges for the Future." At the hearing, former surgeon general Dr. Richard Carmona testified about his experiences working within the administration of President George W. Bush and the resistance he experienced in attempting to carry out what he perceived to be his duties in that office. A segment of that hearing is reprinted here. (Ellipses, . . ., indicate the omission of certain passages from the hearing record.)

Chairman WAXMAN. . . .

You came in as Surgeon General in 2002, and at that time there was a great national debate about the role of stem cells in medical research. I understand you thought the Surgeon General could play a constructive role in explaining this issue, just the science of it, to the American public. Could you tell us what you tried to do, and what the result was?

Dr. CARMONA. Yes, Mr. Chairman, I would be happy to.

I recognize that notwithstanding stem cell issues, the Nation suffers from health illiteracy. The literature is clear, about a third of the Nation really doesn't understand the science we have to deal with every day, it doesn't understand the relationship that their behavior has to ultimate health outcomes. And I saw this debate going around not only as a Surgeon General, but I witnessed it as a professor, and I saw that much of the discussion was being moved forward devoid of science.

And so I approached leadership to say the Surgeon General should be leaning forward on this; we should be, in fact, in the debate on this issue so that we make sure the American public, and our elected officials, our appointed officials are all knowledgeable of the science.

Much of the discussion was being driven by theology, ideology, and preconceived beliefs that were scientifically incorrect. So I thought this is a perfect example of the Surgeon General being able to step forward, educate the American public as well as elected appointed officials so that we can have, if you will, informed consent on an issue to the American public to make better decisions.

I was blocked at every turn. I was told the decision had already been made, stand down, don't talk about it. That information was removed from my speeches.

Chairman WAXMAN. Who would remove a portion of your speech?

Dr. CARMONA. There were people who were actually assigned in the Department to vet my speeches to speechwriters who were helping me put together talking points and things like that. Unfortunately I was naive enough during my first year that I didn't recognize this was happening. Many of the staff, in trying to protect me, didn't tell me the embattled problems and positions that they were in in trying to help me bring the best science forward, but constantly being vetted, and politically vetted, I should say, not scientifically vetted. And it was a while before I figured out that this was happening behind the scenes.

Chairman WAXMAN. Did you have any of your other speeches vetted and censored?

Dr. CARMONA. Repeatedly.

Chairman WAXMAN. Repeatedly[?]

Dr. CARMONA. Yes.

Chairman WAXMAN. And were these scientists or physicians that were doing it, or political people?

Dr. CARMONA. No. In fact, I welcome input from my colleagues on science. I often called my NIH colleagues and CDC, my officers in other departments, to say, what do you think about this, give me the best science. And I would bring groups together to achieve consensus on a scientific issue.

The vetting was done by political appointees who were specifically there to be able to spin, if you will, my words in such a way that would be preferable to a political or ideologically preconceived notion that had nothing to do with science.

Chairman WAXMAN. Were you allowed to speak freely to reporters?

Dr. CARMONA. No. I was often instructed what to say or what not to say. I did the best I could to speak out on issues honestly. I never lied, I never covered the truth. But it was a fine line that I walked all the time, because often the particular issue already had a preconceived political solution, and I had nothing to do with it. And what I found in my first year was that I would see policy moving forward, and I would scratch my head and think, shouldn't the Surgeon General have been involved in this discussion? Yet I had nothing to do with it, but yet be expected to support these notions that were released to the press, through policy, legislation and such. I had no input into them prospectively.

Chairman WAXMAN. The President made a decision on stopping research using embryonic stem cells. He claimed he had a certain number of lines of cells that were already in existence, and he would allow that research to go forward. It may not have been the decision you agreed with, but it was his decision. What do you think your role should be after the President decides for the administration what that administration's policy would be?

Dr. CARMONA. Well, Mr. Chairman, I think clearly the President of the United States, as the senior elected official, has the authority to do what he sees fit, as does Congress as the elected officials representing our citizens. However, I think as part of the due diligence the Surgeon General should be at the table representing our colleagues in science as it relates to the issue.

Source: The Surgeon General's Vital Mission: Challenges for the Future. Hearing before the Committee on Oversight and Government Reform. House of Representatives. One Hundred Tenth Congress. First Session. July 10, 2007.

https://house.resource.org/110/gov.house.ogr.20070710_
hrs15REF2154.pdf. Accessed on July 24, 2013.

_____

In the Matter of Julie MacDonald (2008)

In November 2007, Senator Ron Wyden (D-Ore.) asked the U.S. Department of the Interior's (DOI) Office of Inspector General (OIG) to investigate decisions made about 18 endangered species by Deputy Assistant Secretary for Fish, Wildlife and Parks, Julie MacDonald. Wyden had reason to believe that MacDonald's decisions had been made on the basis of political, rather than scientific, factors. In March 2007, the OIG released its findings on the matter. In December 2008, it produced a longer and more detailed report on the MacDonald matter, in which it summarized its 141-page report as follows:

The nature and extent of MacDonald's influence varied greatly. For example, in one instance we found that MacDonald went to extraordinary efforts to steer a particular decision, but ultimately her efforts had no effect on the outcome. In other instances, her involvement clearly caused a particular result to occur. Ironically, in several instances she played no role in the decision-making process, but because of her reputation FWS personnel believed she had, in fact, been exerting influence. One FWS employee told us that MacDonald's influence was so prevalent that "it became a verb for us—getting MacDonalded."

We reaffirmed findings from previous OIG investigations which showed that MacDonald pursued her agenda by exerting political influence on the FWS Washington Office, regional offices, and field offices. She frequently contested the scientific findings of FWS biologists and often replaced their scientific conclusions with her own, even though she was not a biologist. MacDonald also acted as an economist–again without professional training—in her efforts to restrict critical habitat designation (CHD). In fact, her attempts to perform an analysis of the economic impact of one particular CHD resulted in "math errors" of "an order of magnitude" that led to the exclusion of critical habitat from the rule published in the *Federal Register*. According to FWS personnel, the agency spent approximately $100,000 to republish a corrected version of the rule.

MacDonald's zeal to foster her agenda caused significant harm to the integrity of the ESA [Endangered Species Act] decision-making process, along with potential harm to FWS' reputation among its state and local sister agencies. Moreover, MacDonald's actions resulted in the untold waste of hundreds of thousands of taxpayer dollars in the form of unnecessary litigation costs defending lawsuits, as well as those costs associated with redoing ESA decisions mandated by the courts. Indeed, MacDonald's attempts to manipulate the "best available science" was not lost on the federal courts; in its ruling overturning FWS' greater sage grouse decision, the U.S. District Court for the District of Idaho observed, "MacDonald's principal tactic is to steer the 'best science' to a pre-ordained outcome . . ." The court concluded, "For that reason, MacDonald's extensive involvement in the sage-grouse decision is an independent reason for the Court's finding that the Director's 12-Month Finding is arbitrary and capricious. . . ."

Finally, we found that many FWS employees believe their daily work continues to be hampered by the lack of clear and established policies for implementation of the ESA. In instances where policies do exist, they appeared to change from listing decision to listing decision. One employee told us he would wake in the morning and ask, "Okay, what's the agency doing today?" This, he concluded, was "a problem." MacDonald was clearly able to use these policy voids to impose her will on the ESA process.

MacDonald resigned her post on May 1, 2007, because, she said, of "public pressure."

Source: *Investigative Report: The Endangered Species Act and the Conflict between Science and Policy.* Office of the Inspector General. Department of the Interior. http://www.slvec.org/

**Interference at the EPA: Science and Politics at the U.S.
Environmental Protection Agency (2008)**

*In 2007, the Union of Concerned Scientists (UCS), having heard
of a number of cases of political interference in the work of scientists
at the U.S. Environmental Protection Agency (EPA), carried
out a survey of 5,500 scientists who worked for the organization,
asking a series of 44 questions about political interference in their
own work or of which they had heard. Based on this survey, the
UCS made a number of recommendations for reducing this problem
in the future. Those recommendations are as follows.*

Recommendations

The many forms of political interference in EPA science revealed through our survey, our interviews, and other sources
of information require a suite of solutions in five major are-
nas: protecting EPA scientists, increasing agency transparency,
reforming its regulatory process, strengthening its scientific
advisory system, and depoliticizing funding, monitoring, and
enforcement.

- **Protecting EPA Scientists**: The agency's scientists have a
  profound responsibility to the U.S. public. To fulfill that
  responsibility, they need reassurance that standing behind
  their scientific work will not open them to official or unof-
  ficial retaliation. Congress is considering several bills that
  would strengthen the federal whistle-blower system. Congress should pass the strongest possible protections, and the
  next EPA administrator should formally incorporate them
  into the agency's policies.
- **Making the EPA More Transparent**: Decisions made behind closed doors threaten the integrity of EPA science and

the agency's ability to protect public health and the environment. Opening up these decisions to congressional and public scrutiny is an important step in revealing and ending the misuse of science. The EPA should institute a transparency policy for all meetings with representatives of other federal agencies and outside entities. The agency should also create procedures that ensure the periodic release of scientific documents and prevent them from remaining in draft form indefinitely. The EPA should publish a summary statement discussing the scientific basis for each significant regulatory decision, and document dissenting opinions. The agency should also reform its policies to allow scientists to communicate freely with the media, and to quickly clear their findings for publication in scientific journals, to ensure the free flow of scientific information.

- **Reforming the Regulatory Process**: The EPA was created to implement and enforce the nation's environmental laws, and it has developed the expertise, experience, processes, and policies needed to fulfill that charge. While the White House is responsible for overseeing federal agencies, it must strike a better balance between administration priorities and agency independence. The White House should respect the agency's reservoir of scientific and technical knowledge and restrain the OMB from reviewing the EPA's scientific and technical documents. To ensure the central role of the environment in high-level decision making, the next president should elevate the EPA to a cabinet-level agency. Congress should also consider how to reform and strengthen our nation's regulatory structure, to meet the pressing environmental challenges of the twenty-first century.

- **Ensuring Robust Scientific Input to the EPA's Decision Making**: The EPA should review and strengthen how it uses the scientific expertise of its staff and external advisory committees to create policies—especially when scientific input is critical or required by law. Specifically, the next EPA administrator should work with the Clean Air Science

Advisory Committee to improve the process for setting the National Ambient Air Quality Standards, to ensure that the administrator relies on the "best available science." The agency should also tighten its conflict-of-interest restrictions.

- **Depoliticizing Funding, Monitoring, and Enforcement**: Problems with funding, monitoring and enforcement also need to be addressed by Congress and the next president to ensure that the EPA is the robust environmental agency that our country needs. Congress should provide the EPA with resources commensurate with its growing responsibilities and should work to ensure that selective internal budget cuts are not used to punish inconvenient programs or offices. The next president should commit to strong and consistent enforcement of the nation's environmental laws.

Source: Donaghy, Timothy, Francesca Grifo, and Meredith McCarthy. *Interference at the EPA: Science and Politics at the U.S. Environmental Protection Agency.* Cambridge, MA: Union of Concerned Scientists, 2008, 7–8. Reprinted by permission.

State of Louisiana, Act No. 473, Regular Session 2008

For at least two decades, the State of Louisiana has been exploring methods of challenging the teaching of organic evolution in its public schools. In 1981, for example, it adopted the Balanced Treatment for Creation-Science and Evolution-Science in Public School Instruction Act, requiring all public schools in the state offer instruction in creation science along with the teaching of evolution in biology classes. That law was declared unconstitutional by the U.S. Supreme Court in 1987. The state's latest effort in this direction occurred in 2008, when the legislature adopted a much broader act, allowing public school teachers to use materials not normally part of the regular curriculum to be used in discussions as to the possible errors in a number of otherwise widely held scientific

theories. The law was promoted as an effort to improve students'
skills of critical thinking.

An Act

To enact R.S. 17:285.1, relative to curriculum and instruction; to provide relative to the teaching of scientific subjects in public elementary and secondary schools; to promote students' critical thinking skills and open discussion of scientific theories; to provide relative to support and guidance for teachers; to provide relative to textbooks and instructional materials; to provide for rules and regulations; to provide for effectiveness; and to provide for related matters.

Be it enacted by the Legislature of Louisiana:

Section 1. R.S. 17:285.1 is hereby enacted to read as follows:

§285.1. Science education; development of critical thinking skills

A. This Section shall be known and may be cited as the "Louisiana Science Education Act."

B.(1) The State Board of Elementary and Secondary Education, upon request of a city, parish, or other local public school board, shall allow and assist teachers, principals, and other school administrators to create and foster an environment within public elementary and secondary schools that promotes critical thinking skills, logical analysis, and open and objective discussion of scientific theories being studied including, but not limited to, evolution, the origins of life, global warming, and human cloning.

(2) Such assistance shall include support and guidance for teachers regarding effective ways to help students understand, analyze, critique, and objectively review scientific theories being studied, including those enumerated in Paragraph (1) of this Subsection.

C. A teacher shall teach the material presented in the standard textbook supplied by the school system and thereafter may use supplemental textbooks and other instructional materials to help students understand, analyze, critique, and review scientific theories in an objective manner, as permitted by the city, parish, or other local public school board unless otherwise prohibited by the State Board of Elementary and Secondary Education.

D. This Section shall not be construed to promote any religious doctrine, promote discrimination for or against a particular set of religious beliefs, or promote discrimination for or against religion or nonreligion.

E. The State Board of Elementary and Secondary Education and each city, parish, or other local public school board shall adopt and promulgate the rules and regulations necessary to implement the provisions of this Section prior to the beginning of the 2008–2009 school year.

Section 2. *[Specifies conditions under which the act becomes law.]*

Source: Act No. 473. http://www.legis.la.gov/Legis/View-Document.aspx?d=503483&n=SB733%20Act%20473. Accessed on July 23, 2013.

Proposed Amendments to H.R. 1 (2011)

H.R. 1 was a continuing resolution designed to fund the U.S. government through September 30, 2011. Rep. Darrell Issa (R-Calif.) offered a number of amendments to that act, reflecting his view that certain types of scientific research should not be funded by the U.S. government. Those amendments were as follows:

AMENDMENT NO. 417: At the end of the bill (before the short title), insert the following:

Sec. __. None of the funds made available by this Act may be used by the National Institutes of Health to study the impact of integral yoga on hot flashes in menopausal women.

AMENDMENT NO. 418: At the end of the bill (before the short title), insert the following:

Sec. __. None of the funds made available by this Act may be used by the National Institutes of Health to examine the potential impact of a soda tax on population health.

AMENDMENT NO. 419: At the end of the bill (before the short title), insert the following:

Sec. __. None of the funds made available by this Act may be used by the National Institutes of Health to research the use of marijuana in conjunction with opioid medications, such as morphine.

AMENDMENT NO. 420: At the end of the bill (before the short title), insert the following:

Sec. __. None of the funds made available by this Act may be used by the Department of Health and Human Services to study condom use skills in adult males.

AMENDMENT NO. 421: At the end of the bill (before the short title), insert the following:

Sec. __. None of the funds made available by this Act may be used by the Department of Health and Human Services to study the concurrent and separate use of malt liquor and marijuana among young adults.

AMENDMENT NO. 422: At the end of the bill (before the short title), insert the following:

Sec. __. None of the funds made available by this Act may be used by the National Science Foundation to study whether video games improve mental health for the elderly.

Source: "Issa Amendments to H.R. 1." Coalition to Promote Research. http://www.cossa.org/CPR/2011/Issa.pdf. Accessed on May 6, 2013. All of these amendments were withdrawn by Representative Issa before being voted upon. See http://rsc.scalise.house.gov/uploadedfiles/rsc_member_ health_care_initiatives_in_the_112th_congress.pdf. Accessed on May 6, 2013.

_____

Allegation of Scientific and Scholarly Misconduct and Reprisal (2012)

The Klamath River system that runs through Oregon and California was once the third-largest center of salmon run on the West Coast of the United States. By the beginning of the twenty-first century, salmon runs had been reduced on the river to a level much lower than that of previous decades. In an effort to restore salmon runs to their historic levels, a group of farmers, commercial fishermen, environmentalists, politicians, Native Americas, and others joined together to develop a plan for the removal of a number of dams on the rivers that were impeded salmon runs on the river. During the process of assessing the validity of this plan, a question was raised by one of the researchers involved as to the impartiality of the Department of the Interior about the environmental research being conducted. One researcher, Dr. Paul Houser, eventually came to the conclusion that the department was ignoring important relevant data and that, when he raised questions about this point, he was subjected to reprisals from the department. A portion of the complaint he filed to the department appears here. Ellipses (. . .) represent omitted sections of the document.

ALLEGATION I: Intentional Falsification: Motivated by Secretary Salazar's publically stated intention to issue a Secretarial determination in favor of removing four dams on the Klamath River (due on March 31, 2012), the Department of the Interior has likely followed a course of action to construct such an outcome. In 2009, Secretary Salazar stated that the proposal to remove the Klamath River dams "will not fail," and on September 19, 2011, Ms. Kira Finkler, Deputy

Commissioner for External and Intergovernmental Affairs, told me directly that "the Secretary wants to remove those dams." This intention has motivated Department of the Interior officials to "spin" or incompletely report the scientific results towards a more optimistic scientific story that supports dam removal. . . .

ALLEGATION II: Intentionally circumventing policy that ensures the integrity of science and scholarship, and actions that compromise scientific and scholarly integrity. . . .

On September 15, 2011, I expressed concern via written disclosure relating to the scientific integrity of a draft press release on the draft environmental analysis for removing four Klamath River dams, and via verbal disclosure about the integrity of the larger Klamath River dam removal Secretarial determination process. My disclosure was never directly addressed, and supervisors have enacted and used 1-year probationary status to enact reprisal culminating in the termination of my employment (effective February 24, 2011). The details leading to the termination show a pattern of hindering and not being supportive or honest about the scientific integrity process; the details themselves are not the scientific integrity issue but are rather a case of subsequent reprisal that show intentional actions that compromise the scientific and scholarly integrity codes called out above. Below I outline the actual disclosure, and subsequent reprisal. . . .

In response to this complaint, the Department of the Interior convened a three-member panel of experts to assess the validity of these claims. On the basis of the panel's review, the department determined that both of the complaints were "Not Warranted." The department did conclude its report with the following note:

In conclusion, through the evaluation of your allegations according to 305 DM 3, and measured against the requirements for finding of misconduct as defined in Section 3.5.M(3), I found no merit in the charges. The allegations presented in this complaint are considered closed.

The Department and Dr. Houser later issued a brief joint statement about the resolution of his complaint:

"Dr. Houser and the Bureau of Reclamation have jointly agreed to resolve his complaint filed under the Whistleblower Protection Act through the U.S. Office of Special Counsel's Alternative Dispute Mediation Program to the mutual satisfaction and the best interest of both parties."

This statement constitutes the only comments either party shall make on this specific matter.

Sources: [Untitled document; letter from Dr. Houser to Interior]. http://www.klamathbasincrisis.org/settlement/science/houserallegation022412.pdf. Accessed on May 9, 2013; [Untitled document; letter from Interior to Dr. Houser]. http://www.peer.org/assets/docs/doi/3_25_13_Interior_adoption.pdf. Accessed on May 9, 2013; "Scientific Whistleblower Complaint Resolved." Protecting Employees Who Protect Our Environment. http://www.peer.org/news/news-releases/2012/12/04/scientific-whistleblower-complaint-resolved/. Accessed on May 9, 2013.

Annie Tummino, et al.,—against—Margaret Hamburg, Commissioner of Food and Drugs, et al., No. 12-cv-763 (ERK)(VVP) (2013)

The use of high-dose hormone pills to prevent conception after sexual intercourse has been studied for a half century. Such pills are known as emergency contraceptives or, more commonly, morning-after pills. The first morning-after pills became commercially available in the United Kingdom in October 1984, and in September 1998, the U.S. Food and Drug Administration (FDA) approved the first such pill in the United States. Ever since the pill was first introduced in the United States, many health care advocates have recommended that the pill be made available without prescription to females of all ages. In February 2006, the FDA authorized the nonprescription sale of the pill to women over the age of 18. Since that time, some health care advocates have continued to press for the nonprescription sale of the pill even for females under the age of 18. In April 2013, Judge Edward R. Korman of the U.S. District Court for the Eastern District of New York issued a memorandum requiring the federal government to take just such an action, allowing nonprescription sale of the drug to all females. Some major portions of that memorandum are as follows (omitted portions are indicated by ellipses (. . .):

This action was originally brought in January 2005 to challenge the FDA's denial of a Citizen Petition seeking over-the-counter ("OTC") access to Plan B for women of all ages. The complaint asserted that the FDA's denial of the Citizen Petition, which it considered along with a number of proposals regarding over-the-counter access to emergency contraception submitted by Plan B's sponsor, was arbitrary and capricious because it was not the result of reasoned and good faith agency decision-making. In a prior opinion, I concluded that the plaintiffs were right. Tummino v. Torti, 603 F. Supp. 2d 519 (E.D.N.Y. 2009). . . .

This case is not about the potential misuse of Plan B by 11-year-olds. These emergency contraceptives would be among the safest drugs sold over-the-counter, the number of 11-year-olds using these drugs is likely to be miniscule, the FDA permits drugs that it has found to be unsafe for the pediatric population to be sold over-the-counter subject only to labeling restrictions, and its point-of-sale restriction on this safe drug is likewise inconsistent with its policy and the Food, Drug, and Cosmetic Act as it has been construed. Instead, the invocation of the adverse effect of Plan B on 11-year-olds is an excuse to deprive the overwhelming majority of women of their right to obtain contraceptives without unjustified and burdensome restrictions. . . .

In my 2009 opinion, I traced the evidence demonstrating that the conduct of the FDA was influenced by the Bush White House, acting through the Office of the Commissioner of the

FDA, and I held that this kind of political interference called into serious question the legitimacy of the FDA's decision. In the present circumstances, the political interference came directly from the Secretary of Health and Human Services, a member of the President's Cabinet. . . .

The motivation for the Secretary's action was obviously political. "It was the first time a cabinet member had ever publicly countermanded a determination by the F.D.A., the agency charged with ensuring the safety of foods and medicines." Gardiner Harris, White House and the FDA Often at Odds, N.Y. Times, Apr. 3, 2012 at A1. And it was an election-year decision that "many public health experts saw as a politically motivated effort to avoid riling religious groups and others opposed to making birth control available to girls."

> **Source:** "Annie Tummino, et al.,—against—Margaret Hamburg, Commissioner of Food and Drugs, et al., No. 12-cv-763 (ERK)(VVP)." https://www.nyed.uscourts.gov/sites/default/files/opinions/Tummino%20SJ%20memo.pdf. Accessed on May 9, 2013. For the original decision in this case, See Annie Tummino et al.,—against—Frank M. Torti, No. 05-CV-366 (ERK)(VVP), https://www.nyed.uscourts.gov/sites/default/files/opinions/05cv366mofinal.pdf. Accessed on May 9, 2013.

———

Draft High Quality Research Act (2013)

For more than a decade, a number of legislators and politicians, most of them members of the Republican Party, have argued for changes in the process of scientific research to make sure that such research is "sound" and of "high quality." An example of the kind of changes recommended by such individuals is a draft bill proposed by Rep. Lamar Smith (R-Tex.), chair of the House Committee on Science, Space, and Technology. Representative Smith's draft bill evoked widespread concern among scientists and their allies, both in the United States and in other parts of the world. Representative of the kinds of concerns expressed about the bill are the comments made in letter to Smith by Rep. Eddie Bernice Johnson (D-Tex.), ranking member of the same committee. The bill and portions of Representative Johnson's letter are reprinted here. (Omitted text is indicated by ellipses, . . .)

Section 1. Short title.

This Act may be cited as the "High Quality Research Act."

Sec. 2. High quality research.

(a) CERTIFICATION.—Prior to making an award of any contract or grant funding for a scientific research project, the Director of the National Science Foundation shall publish a statement on the public website of the Foundation that certifies that the research project—

(1) is in the interests of the United States to advance the national health, prosperity, or welfare, and to secure the national defense by promoting the progress of science;

(2) is the finest quality, is ground breaking, and answers questions or solves problems that are of utmost importance to society at large; and

(3) is not duplicative of other research projects being funded by the Foundation or other Federal science agencies.

(b) TRANSFER OF FUNDS.—Any unobligated funds for projects not meeting the requirements of subsection (a) may be awarded to other scientific research projects that do meet such requirements.

(c) INITIAL IMPLEMENTATION REPORT.—Not later than 60 days after the date of enactment of this Act, the Director shall report to the Committee on Commerce, Science, and Transportation of the Senate and the Committee on Science, Space, and Technology of the House of Representatives on how the requirements set forth in subsection (a) are being implemented.

(d) NATIONAL SCIENCE BOARD IMPLEMENTA-
TION REPORT.—Not later than 1 year after the date of
enactment of this Act, the National Science Board shall report to the Committee on Commerce, Science, and Transportation of the Senate and the Committee on Science,
Space, and Technology of the House of Representatives its
findings and recommendations on how the requirements of
subsection (a) are being implemented.

(e) IMPLEMENTATION BY OTHER AGENCIES.—
Not later than 1 year after the date of enactment of this Act,
the Director of the Office of Science and Technology Policy,
in collaboration with the National Science and Technology
Council, shall report to the Committee on Commerce, Science, and Transportation of the Senate and the Committee
on Science, Space, and Technology of the House of Representatives on how the requirements of subsection (a) may be
implemented in other Federal science agencies.

Source: http://www.cossa.org/CPR/2013/HighQualityRe
searchAct.pdf. Accessed on July 30, 2013.

[Letter from Johnson to Smith]

. . . This is the first step on a path that would destroy the
merit-based review process at NSF and intrudes political pressure into what is widely viewed as the most effective and creative process for awarding research funds in the world. It is this
process that has supported the growth of the American research
university system and it is this process that has established the
American research enterprise as the most innovative of our age.

No system constructed of, for, and by humans is infallible.
But for decades the world has held the NSF's peer review process as the gold standard for how scientific proposals should be
judged and funded. This applies equally to all fields of science,
including the social and behavioral sciences. In this context,
the term "peer" is not simply a fellow citizen as we encounter
on a courtroom jury. It means very specifically another scientist

with expertise in at least some aspect of the science being proposed. Politicians, even a distinguished Chairman of the Committee on Science, Space, and Technology, cannot be "peers" in any meaningful sense.

. . .

Interventions in grant awards by political figures with agendas, biases, and no expertise is the antithesis of the peer review processes. By making this request, you are sending a chilling message to the entire scientific community that peer review may always be trumped by political review. You also threaten to compromise the anonymity that is crucial to the frank and open exchange of comments and critiques during the review process, and in doing so, further compromise the integrity of the merit review process. How can future participants in the peer review process have confidence that their work will remain confidential when the Chairman of the Science, Space, and Technology Committee has shown that probing specific award absent any allegation of wrong-doing may become the way business is done?

Source: Letter from Rep. Eddie Bernice Johnson to Lamar Smith. http://big.assets.huffingtonpost.com/42613Johnson SmithNSFgrants.pdf. Accessed on July 30, 2013.

FRESH

Introduction

Scholars and interested observers have been writing about the relationship of science and politics ever since science became important in the policy-making activities of a government. This chapter lists some of the most interesting and most important books, articles, reports, and Internet web pages dealing with this topic. In some cases, an article may have appeared in both print and electronic forms, in which case it is listed in the most easily accessible of those forms.

Books

Badash, Lawrence. 2009. *A Nuclear Winter's Tale: Science and Politics in the 1980s.* Cambridge, MA: MIT Press.

> A theory was put forward in the 1980s that dust and debris produced in a number of nuclear explosions could be so intense and extensive that they would block out the Sun's rays for years, producing an extended "nuclear winter" throughout the world. The author discusses the ways in which this concept corresponded with political

---

A plant physiologist compares Florida-grown Endless Summer tomatoes with his greenhouse-grown fruit. All contain bioengineered genes. The increased availability of genetically modified fruits and vegetables has spurred concerns over health and environmental safety. (Agricultural Research Service/USDA)

objectives in the United States during the Cold War with the Soviet Union and shows how scientific research was affected by political objectives with regard to the issue.

Berezow, Alex B., and Hank Campbell. 2012. *Science Left Behind: Feel-good Fallacies and the Rise of the Anti-scientific Left.* New York: Public Affairs.

> The authors argue that conservatives are frequently criticized for ignoring consensus in the scientific community about issues such as evolution and climate change, but that critics ignore the tendency of many liberals also to reject ideas with which they are not comfortable, such as significant bodies of research on alternative forms of energy, basic biological research, and immunization practices.

Bird, Kai, and Martin J. Sherwin. 2005. *American Prometheus: The Triumph and Tragedy of J. Robert Oppenheimer.* New York: A. A. Knopf.

> This biography of one of the leaders of the Manhattan Project during World War II to build a nuclear weapon illustrates how closely intertwined scientific and political issues were during this period, and how Oppenheimer's refusal to support construction of a fusion weapon resulted in his being ostracized from the halls of power late in his life.

Bolin, Bert. 2007. *A History of the Science and Politics of Climate Change: The Role of the Intergovernmental Panel on Climate Change.* Cambridge, UK: Cambridge University Press.

> The first chairman of the Intergovernmental Panel on Climate Change provides an excellent review of the ongoing interaction between scientists who were conducting research on global climate change and their critics from politics, government, and the general population.

Clark, William R. 2008. *Bracing for Armageddon?: The Science and Politics of Bioterrorism in America.* New York: Oxford University Press.

This book focuses on a relatively new aspect of the interaction of science and politics, issues that arise as a result
of the spread of bioterrorism. The author reviews the science and politics related to the topic and discusses ways in
which the two fields interact in both positive and negative
ways.

DeGette, Diana, and Daniel Paisner. 2008. *Sex, Science, and
Stem Cells: Inside the Right Wing Assault on Reason.* Guilford,
CT: Lyons Press.

DeGette is the Democratic representative from the First
District of Colorado. She and the coauthor present a
strong condemnation of what they see as an attack on
the scientific process by members of the administration of
President George W. Bush during his eight years in office.

Dupré, J. Stefan, and Sanford A. Lakoff. 1992. Englewood
Cliffs, NJ: Prentice-Hall, Inc.

The authors discuss a number of issues involved in the
relationship between the scientific establishment and the
federal government in the years following World War II.

Farrington, Benjamin. 1939. *Science and Politics in the Ancient
World.* New York: Barnes & Noble, Inc.

A remarkable book in which a highly regarded historian
of science discusses the controversy between science and
government in the last few centuries of the pre–Christian
era. The book is of special interest because so much of the
controversy that Farrington describes has reappeared on a
number of occasions in more recent history.

Ferris, Timothy. 2010. *The Science of Liberty: Democracy, Reason, and the Laws of Nature.* New York: Harper.

Many scholars have argued that the development of a
scientific community and a scientific method for solving
problems is dependent upon a society which permits and
encourages free and independent thought. Ferris posits

the quite different notion that the reverse is true, and it is the growth of the scientific way of thinking about the world that has contributed to the growth and development of democratic institutions and states around the world over the last three centuries.

Findel, Madelon Lubin. 2007. *Truth, Lies, and Public Health: How We Are Affected When Science and Politics Collide*. Westport, CT: Praeger.

The author explores the myriad ways in which public policy and action in the field of public health are affected by the interaction of science and politics. She focuses on the fields of contraception, the HIV/AIDS epidemic, the stem cell controversy, control of marijuana, needle exchange programs for the control of HIV, tuberculosis control, dietary supplements, silicone breast implantation, and immunization against infectious diseases.

Foerstel, Herbert N. 2010. *Toxic Mix?: A Handbook of Science and Politics*. Santa Barbara, CA: Greenwood Press/ABC-CLIO.

The author reviews controversies that have arisen in a number of fields when politics becomes embroiled with science, including the HIV/AIDS epidemic, nuclear energy, space science, public health policy, stem cell research, environmental issues, climate change, evolution, and human reproduction.

Goodman, Jordan, and Vivien Walsh. 2001. *The Story of Taxol: Nature and Politics in the Pursuit of an Anti-cancer Drug*. Cambridge, UK; New York: Cambridge University Press.

Taxol was arguably the most important natural product drug discovered at the end of the 20th century. Its efficacy in the treatment of cancer made its discovery, synthesis, production, and use and event of enormous importance that produced a number of political and social effects which, in turn, had their own effects on the later history of the drug.

Gough, Michael, ed. 2003. *Politicizing Science: The Alchemy of Policymaking*. Stanford, CA: Hoover Institution Press; Washington, DC: George C. Marshall Institute.

The papers that make up this book discuss the harms that can come about as the result of manipulating science for the purpose of political gain. It differs from a number of other books in this bibliography because it presents an analysis from the conservative political view, dealing with issues such as the perceived overregulation of business and industry, the manipulation of medical and scientific definitions for political purposes, and the suppression of scientific data to support liberal and progressive causes and views.

Harrison, Neil E., and Gary C. Bryner, eds. 2004. *Science and Politics in the International Environment*. Lanham, MD: Rowman & Littlefield Publishers.

The contributors to this collection of essays remind us that the interaction of science and politics is not strictly an American problem, but one that is present in many societies around the world. Various essays in the book deal with transboundary issues, global climate change, multinational forests, acid rain, and fishing disputes.

Josephson, Paul R. 1991. *Physics and Politics in Revolutionary Russia*. Berkeley, CA: University of California Press.

The author explores the ways in which political doctrine pervasive in the Soviet Union following World War II affected the way scientists carried out their research, having to fit their work into the philosophy of dialectic materialism.

Josephson, Paul R. 1996/2005. *Totalitarian Science and Technology*, 2nd ed. Amherst, NY: Humanities Press.

In the first edition of this book, the author explored the ways in which totalitarian regimes, especially those of Adolf Hitler and Joseph Stalin, directed the work of scientists and engineers to achieve the greater goals of the political parties in control at the time. In this second edition, the author expands his analysis to include modern-day totalitarian states, including China, Cuba, and North Korea. This is an invaluable source of information on the impact of political systems on the direction and character of scientific research.

Keller, Ann Campbell. 2009. *Science in Environmental Policy: The Politics of Objective Advice*. Cambridge, MA: MIT Press.

The author explores the way in which scientists are involved in the process of policy making, with special attention to acid rain and climate change issues. She finds that scientists are involved in policy issues at three points: setting the agenda, legislation, and implementation, the first being the most crucial for all parties involved.

Lakoff, Sanford, and Herbert F. York. 1989. *A Shield in Space? Technology, Politics, and the Strategic Defense Initiative*. Berkeley, CA: University of California Press.

The authors review the creation and progress of the strategic defense initiative, pointing out that it was probably doomed to failure from the outset because it was based on "political rather than scientific judgments."

Meiners, Roger, Pierre Desrochers, and Andrew P. Morriss, eds. 2012. *Silent Spring at 50: The False Crises of Rachel Carson*. Washington, DC: Cato Institute.

The essays in this collection take a look back at one of the most famous books in the history of the modern environmental movement. They find that, given the admitted benefits that flowed from the book, much of its argument was based on questionable, if not actually false, science.

Mooney, Chris. 2012. *The Republican Brain: The Science of Why They Deny Science and Reality*. Hoboken, NJ: Wiley.

The author follows up on updates, and expands on his 2005 book about patterns among Republican Party members involving their denial of information about which there is wide agreement in the scientific community, and how this affects the process of policy decision making in the government.

Mooney, Chris. 2005. *The Republican War on Science*. New York: Basic Books.

Mooney was the lead author of a report released by the Union of Concerned Scientists in 2004 about the troubling practice of members of the administration of President George W. Bush interfering in the normal process by which scientific research is funded and carried out in the United States, and in the ways it is used to determine public policy on a variety of important social issues.

Oreskes, Naomi, and Erik M. Conway. 2010. *Merchants of Doubt: How a Handful of Scientists Obscured the Truth on Issues from Tobacco Smoke to Global Warming*. New York: Bloomsbury Press.

The authors review six instances in which a relatively small number of individuals argued, with substantial success, against viewpoints that a very large proportion of research scientists held on a topic. The issues were the development of strategic defense systems, acid rain, depletion of the ozone layer, the risks posed by secondhand smoke, global climate change, and the contributions of Rachel Carson in her book, *Silent Spring*.

Otto, Shawn Lawrence. 2011. *Fool Me Twice: Fighting the Assault on Science in America*. New York: Rodale.

The author reviews a number of areas in which politicians, government officials, and members of the mass media withhold or misrepresent scientific information that could be used to solve many national problems. He goes on to recommend actions that the average citizen can take to remedy this situation.

Price, Don Krasher. 1954. *Government and Science: Their Dynamic Relation in American Democracy.* New York: New York University Press.

> The author served in a number of advisory positions in the U.S. government, and this series of lectures discusses his views on the relationship between the scientific community and the federal government in the post–World War II years.

Resnick, David B. 2009. *Playing Politics with Science: Balancing Scientific Independence and Government Oversight.* New York: Oxford University Press.

> Resnick reviews some of the general principles involved in the conflict between science and politics that appeared to reach a high level during the first decade of the 21st century. He writes about the autonomy of science, providing government with scientific advice, government funding of science, science and national security, protecting human subjects of research, and education in science.

Roll-Hansen, Nils. 2005. *The Lysenko Effect: The Politics of Science.* Amherst, NY: Humanity Books.

> This book provides an excellent overview of the impact that political factors, as represented by the untrained "barefoot professor," T. D. Lysenko, had on agriculture research and practice during the mid-20th century.

Shilts, Randy. 1987. *And the Band Played on: Politics, People, and the AIDS Epidemic.* New York: St. Martin's Press, 1987.

> In one of the most important books written about the HIV/AIDS epidemic, Shilts discusses in detail the way in which political believes affected the attention paid to the epidemic and the care (or, more accurately, the lack of attention and care) paid to those who were infected with the HIV virus.

Shulman, Seth. 2004. *Scientific Integrity in Policymaking: An Investigation into the Bush Administration's Misuse of Science.* Cambridge, MA: Union of Concerned Scientists.

Snyder, Solomon H. 1989. *Brainstorming: The Science and Politics of Opiate Research*. Cambridge, MA: Harvard University Press.

> The author describes his work on the discovery of opiate receptors in the 1970s and the way that researcher was influenced and affected by a host of social and political factors, not the least of which was President Richard M. Nixon's "War on Drugs" that had been "declared" in 1971.

Solomon, Joan, ed. 1987. *Science in a Social Context*. Oxford, UK: Blackwell Publishing.

> This is a wonderful series of eight books covering topics such as the sociology of science; research and technology as economic activities; and science, technology, and the modern industrial state, with a good deal of information about the interrelationship of science politics.

Sunstein, Cass R. 2009. *Going to Extremes: How Like Minds Unite and Divide*. Oxford, UK; New York: Oxford University Press.

> The author reviews an extensive body of research that shows that bringing people of similar convictions together tends to strengthen common beliefs among those who make up the group. He shows how this tendency contributes to the politicization of views on scientific issues.

Szasz, Thomas. 2001. *Pharmacracy: Medicine and Politics in America*. Westport, CT: Praeger.

> Psychiatrist Szasz is a long-time critic of the use of psychology, psychiatry, psychoanalysis, and related fields to create scientifically "valid" categories of human behavior not based on subjective research, but on certain social and political preferences. For a complete list of his works in this field, see "The Writings of Thomas S. Szasz, M.D.," at http://www.szasz.com/publist.html.

Teller, Edward, and Judith L. Shoolery. 2001. *Memoirs: A Twentieth-century Journey in Science and Politics*. Cambridge, MA: Perseus Publishers.

Teller was one of the great contributors to the development of nuclear science in the last half of the 20th century, but he was often in the news because of his conservative political opinions and the effects those opinions had on both his research recommendations and his political statements.

Tucker, William H. 1994. *The Science and Politics of Racial Research*. Urbana, IL: University of Illinois Press.

Studies of the innate differences between ethnic and racial groups have been popular fields of research for many years. The author argues that such studies have no inherent value as research topics, and that they are carried out for the purpose of justifying policies that create and support social, political, and economic policies of discrimination against certain racial and ethnic groups.

Weiss, Sheila Faith. 2010. *The Nazi Symbiosis: Human Genetics and Politics in the Third Reich*. Chicago; London: The University of Chicago Press.

The 1930s and 1940s was a period that saw an almost unique collaboration between geneticists and the Nazi party in Germany, with the political philosophy of the party providing unmatched opportunities for research in genetic and eugenics that professional scientists found difficult to refuse. The result of this collaboration was the conduct of some of the most heinous scientific studies in the field that have ever been carried out.

Articles

A number of scientific journals offer columns and stories on either a regular or irregular basis. The journal *Nature*, for example, has a bimonthly column called "Worldview," in which columnists discuss specific science/politics issues in many nations of the world. In October and November 2007, the journal also offered a nine-part series of essays by advisers on science policy from Sweden, the United Kingdom, and the United States. A number of journals are devoted exclusively or to a large extent to articles involving the relationship between science and politics. *The Bulletin of the Atomic Scientists* may be the preeminent example of this type of journal, although many others of the same type exist, including *Environmental Science and Policy, Ethics in Science and Environmental Politics, Politics and the Life Sciences,* and *Public Understanding of Science.*

Baggott, Erin. 2006. "A Wealth Deferred—The Politics and Science of Golden Rice." *Harvard International Review.* 28(3): 28–30.

> Golden rice is a genetically engineered product with considerable potential for solving certain health problems common in less developed nations. Yet a number of political factors have prevented the product from wide use. This article describes and discusses the nature of this problem.

Bocking, Stephen. 2013. "Science and Society: The Structures of Scientific Advice." *Global Environmental Politics.* 13(2): 154–159.

> This essay is a review of three books dealing with the making of scientific policy in three geographically diverse regions. The books have in common the recognition that the line between science and politics has become increasingly blurry, with elements from both fields now involved in decisions as to what type of research to carry out and how to do that research. A good general overview of the problems involved in conducting environmental research in the international arena.

Brown, Mark B., and David H. Guston. 2009. "Science, Democracy, and the Right to Research." *Science and Engineering Ethics.* 15(3): 351–366.

> The article explores the question not so much as to whether scientists have the right to conduct research,

but what protection from governmental interference and support by taxpayer dollars they should have in that activity.

Christoforou, Theofanis. 2004. "The Regulation of Genetically Modified Organisms in the European Union: The Interplay of Science, Law and Politics." *Common Market Law Review.* 41(3): 637–709.

> The rapid increase in the number of genetically modified organisms being produced by research has created a host of social, political, and economic issues since the mid-1990s. Christoforou argues that little has been done to try to understand the interwoven issues of science, law, and politics involved in this conflict and offers this article as a first attempt to resolve the problem.

Curry, J. A., P. J. Webster, and G. J. Holland. 2006. "Mixing Politics and Science in Testing the Hypothesis That Greenhouse Warming Is Causing a Global Increase in Hurricane Intensity." *Bulletin of the American Meteorological Society.* 87(8): 1025–1037.

> The authors discuss public reaction to an earlier paper in which they explore possible relationships between the number of hurricanes likely to occur each year in the future and global climate change. They point out that individuals and organizations with little or no background in science seem to receive equal treatment in the media with people who are experts in the area. They pursue the implications of this manner of treating the scientific process and scientific information.

Curtis, Michele G. 2003. "Cloning and Stem Cells: Processes, Politics, and Policy." *Current Women's Health Reports.* 3(6): 492–500.

> The author claims that the science of cloning and stem cell research has become so politicized that it is "almost impossible to develop true consensus on what any national

policy regarding nuclear transfer should look like." She argues, nonetheless, that a dialog over such a policy must take place and the best possible compromise available agreed to.

De Greiff, Alexis. 2006. "The Politics of Noncooperation: The Boycott of the International Centre for Theoretical Physics." *Osiris.* 21(1): 86–109.

In 1974, a controversy developed when the United Nations Education, Scientific, and Cultural Organization (UNESCO) adopted a series of resolutions condemning Israel for its foreign policy and activities, in response to which a group of physicists called for the boycott of the recently established International Centre for Theoretical Physics, an institution created to foster cooperation among physicists from around the world. The author reviews this event and uses it to discuss what it means to "politicize" science.

Dean, Katrina, et al. 2008. "Data in Antarctic Science and Politics." *Social Studies of Science.* 38(4): 571–604.

The authors discuss scientific data collection under terms of the 1961 Antarctic Treaty in which national political factors exerted strong influence as to how the data were collected and shared with other researchers in the field.

Gauchata, Gordon. 2012. "Politicization of Science in the Public Sphere: A Study of Public Trust in the United States, 1974 to 2010." *American Sociological Review.* 77(2): 167–187.

The author examines changes in attitudes (or lack of changes) about trust in science among various groups of individuals. He finds that attitudes toward science have remained relatively stable throughout the period under study for all groups except for conservatives. During the earliest period of the study, conservatives were most trustful of science of all groups' studies, but they were least trustful at the end of the period under study.

Herran, Néstor, and Xavier Roqué. 2013. "An Autarkic Science: Physics, Culture, and Power in Franco's Spain." *Historical Studies in the Natural Sciences.* 43(2): 202–235.

The authors explain how a change of political philosophy in Spain after the ascendancy of Francisco Franco significantly altered the composition and outlook of the country's physics community, attuning it more to the goals and methods of the ruling party.

Howard, Don. 2013. "Einstein's Jewish Science: Physics at the Intersection of Politics and Religion; The Practical Einstein: Experiments, Patents, Inventions." *Physics Today.* 66(1): 42.

Howard reviews two books on the life and work of Albert Einstein that focus on religious, cultural, political, social, and other influences on his work. The question raised in both books is whether there is something specifically "Jewish" or otherwise culturally distinctive about the way Einstein conducted his research. The review article is an excellent introduction to the topic and may lead to an interest in one or both of the two books named in the title of the essay.

Hueston, William D. 2013. "BSE and Variant CJD: Emerging Science, Public Pressure and the Vagaries of Policy-making." *Veterinary Medicine.* 109(3–4): 179–184.

The author reviews the histories of the spread of bovine spongiform encephalopathy (BSE) and Creutzfeld–Jakob disease (CJD), the response by scientific researchers to these problems, and the reaction of political systems and the general population to the new disease threats and draws a number of conclusions about the interaction of science and politics in crises such as these.

Jasen, Patricia. 2005. "Breast Cancer and the Politics of Abortion in the United States." *Medical History.* 49(4): 423–444.

The author reviews the way in which fears about breast cancer have been used by so-called pro-life individuals and groups to frighten women away from having abortions, even after abundant evidence had become available that no such association exists between the two phenomena.

Joffe, Carole. 2013. "The Politicization of Abortion and the Evolution of Abortion Counseling." *American Journal of Public Health*. 103(1): 57–65.

> The author reviews the history of abortion counseling since its origin in the 1970s to the present day and discusses a number of factors that have politicized this practice in clinics across the United States.

Kamerow, Douglas. 2013. "Science Trumps Politics on Emergency Contraception." *BMJ*. 7903: 28.

> The author reviews the controversy over making the so-called morning after pill available to young women of all ages.

Lambright, W. Henry. 2008. "Government and Science: A Troubled, Critical Relationship and What Can Be Done about It." *Public Administration Review*. 68(1): 5–18.

> The author attempts to place the issue of the politicization of science in a historical context and to lay out some suggestions for reducing the conflict present in this relationship in the future.

Logsdon, John M. 1979. "An Apollo Perspective." *Astronautics & Aeronautics*. 17(December 1979): 112–117.

> The author, a highly respected chronicler of space events for many decades, provides an insightful review of the process by which the Apollo program was conceived, developed, and carried out.

Mack, Arien, ed. 2006. "Politics & Science: How Their Interplay Results in Public Policy." *Social Research*. 73(3): whole.

> This issue of the journal is devoted entirely to a collection of articles dealing with specific manifestations of the interaction between science and politics in today's world. Topics include "Science Policy in the United

States: A Commentary on the State of the Art"; "What's New about the Politics of Science?"; "The Politics of Health Care"; "Science, Religion, and the Politics of Stem Cells"; "Abstinence-Only Education: Politics, Science, and Ethics"; "Science and Environmental Policy: The Role of Nongovernmental Organizations"; "Environmental Science Input to Public Policy"; "Can We Still Avoid Dangerous Human-Made Climate Change?"; "Science, Policy, and Politics: Comparing and Contrasting Issues in Energy and the Environment"; and "Climate Change and Nuclear Power." A brief summary of the articles is available at https://epay.newschool.edu/C21120_ustores/web/product_detail.jsp?PRODUCTID=5045.

Manring, Nancy. 1993. "Reconciling Science and Politics in Forest Service Decision Making: New Tools for Public Administrators." *American Review of Public Administration.* 23(4): 343–359.

A number of issues involving Forest Service policy and decisions have involved conflict between the needs of scientific research and political issues. The author explores the effectiveness of using a technique known as alternative dispute resolution (ADR) for dealing with these issues.

Peterson, James C., and Gerald E. Markle. 1979. "Politics and Science in the Laetrile Controversy." *Social Studies of Science.* 9(2): 139–166.

Laetrile is the product of a naturally occurring material, amygdalin, that has been recommended as a cure for cancer by some health practitioners since at least the early 1950s. There has never been any reliable scientific evidence of its efficacy, and considerable evidence of its potential harm to users. This article reviews efforts by proponents of the drug first to obtain scientific validation for its use and when that failed, other approaches for convincing the general public and policy makers of its efficacy.

Pielke, Roger A. Jr. 2004. "When Scientists Politicize Science: Making Sense of Controversy over the Skeptical Environmentalist." *Environmental Science & Policy*. 7: 405–417.

The author offers the hypothesis that scientists themselves are often responsible for the politicization of science by becoming too much involved in the development of public policy, rather than in presenting scientific information by which other individuals are able to do so. He focuses on a 2001 article by Danish statistician Bjørn Lomborg, arguing that environmental problems described by specialists in the field are not really as serious as they are claimed to be.

Reich, M. R. 1983. "Environmental Politics and Science: The Case of Pbb Contamination in Michigan." *American Journal of Public Health*. 73(3): 302–313.

In 1973, between 500 and 1,000 pounds of the highly toxic flame retardant polybrominated biphenyl (PBB) was accidentally shipped to a farmer, who used the product as a food additive for his cattle. Thereafter, the product became widely distributed in the environment, posing a health risk to a large number of humans in the area. This article summarizes the events involved in this incident and analyzes the way that health concerns and scientific issues became politicized as news of the event became public knowledge.

Room, Robin, and Dan I. Lubman. 2010. "Politics and Science in Classifying the Dangers of Drugs." *Evidence Based Mental Health*. 13(4): 97–99.

When new drugs are discovered or introduced, their danger to humans may be classified by governmental or other agencies before the science on which such classifications are made can be fully researched. In such cases, classifications may be based on political, social, economic, or other factors unrelated to science.

Santelli, John S. 2008. "Medical Accuracy in Sexuality Education: Ideology and the Scientific Process." *American Journal of Public Health*. 98(10): 1786–1792.

> The author notes that the Bush administration has been strongly criticized for the unprecedented amount and level of political interference in the scientific process. He explores the validity of this criticism with special attention to federal policies, programs, and practices for the teaching of abstinence-only sex education.

Specter, Michael. 2006. "Political Science: The Bush Administration's War on the Laboratory." *New Yorker*. (March 13, 2006): 58–69.

> The author provides an extensive review of actions taken by various members of the administration of President George W. Bush to interfere with the normal functioning of the scientific process used to aid in the development of public policy on a variety of issues.

Thompson, Kristen M.J., et al. 2013. "Access to Levonorgestrel Emergency Contraception: Science versus Federal Politics." *Women's Health*. 9(2): 139–143.

> The authors suggest that decisions made on the availability of the "morning after" pill for young girls have been based on political, and not scientific, considerations. They explain why this has resulted in an inappropriate decision about use of the medication.

Varzakas, Theodoros H., Ioannis S. Arvanitoyannis, and Haralambos Baltas. 2007. "The Politics and Science behind GMO Acceptance." *Critical Reviews in Food Science and Nutrition*. 47(4): 335–361.

> The introduction of genetically modified foods into the marketplace has been accompanied by extensive debate over their safety and utility in everyday life. This article explores the combination of scientific and political factors involved in this ongoing debate.

Wang, Zuoyue. 1995. "The Politics of Big Science in the Cold War: PSAC and the Funding of SLAC." *Historical Studies in the Physical and Biological Sciences: Journal of the Intellectual and Social History of the Physical Sciences and Experimental Biology since the 17th Century.* 25(2): 329–356.

> As a partial response to the launching of the first man-made satellite, Sputnik, President Dwight D. Eisenhower announced in 1959 that he would be requesting funds from Congress to build the world's largest particle accelerator, the Stanford Linear Accelerator Center (SLAC). This article reviews and discusses the interaction between politics and science in the development of this proposal.

Waxman, Henry A. 2006. "Politics and Science: Reproductive Health." *Health Matrix: The Journal of Law-Medicine.* 16(1): 5–26.

> Congressman Waxman (D-Calif.) reviews information about the politicization of reproductive health issues obtained through research by his Congressional committee. He discusses in some detail the issue with regard to abstinence-only education, abortion and breast cancer, condom use, HIV/AIDS, emergency contraception, and stem cell research.

Weiland, Sabine, Vivien Weiss, and John Turnpenny, eds. 2013. "Special Symposium on Nature, Science and Politics, or, Policy Assessment to Promote Sustainable Development." *Nature + Culture.* 8(1): whole.

> This special issue discusses some of the ways in which science and politics interact with each other in a variety of fields, with special attention to the field of sustainable development. Individual articles in the issue deal with climate change policies in China, old-growth forest conflicts in Lapland, and listing of endangered species in various parts of the world.

Weiss, Robin. 2013. "Sunday Dialogue: Science and Politics." *New York Times*, March 10, 2013, SR2.

> This letter, responses to the letter, and response by the letter writer appeared as part of the newspaper's regular Sunday Dialogue section, in which readers exchange views on important issues of the day, in this case, the politicization of scientific research by politicians and government officials.

Welsh, Rick, and David E. Ervin. 2006. "Precaution as an Approach to Technology Development: The Case of Transgenic Crops." *Science, Technology, & Human Values*. 31(2): 153–172.

> The authors discuss ways in which politicians and researchers from varying political perspectives use the results of scientific research in different ways, and how this practice affects policies and practices developed for certain topics. They use the development of transgenic crops as a case study for this discussion, and suggest some ways in which different viewpoints can more effectively be combined to produce helpful social results.

Willis, Theodore V. 2009. "How Policy, Politics, and Science Shaped a 25-Year Conflict over Alewife in the St. Croix River, New Brunswick-Maine." *American Fisheries Society Symposium*. 69: 793–812. Also available online at http://www.alewifeharvesters .org/wp-content/uploads/2010/03/willis-3-19-09a.pdf. Accessed on August 2, 2013.

> Conflicts between science and politics can occur at every level of government. This article explores such a conflict with regard to the protection of a species of fish on the St. Croix River that flows through New Brunswick and Maine, in which the results of scientific research were unacceptable to some groups of people whose livelihood depended strongly on fishing in the river.

Wise, M. Norton. 2011. "Thoughts on Politicization of Science through Commercialization." *Boston Studies in the Philosophy of Science*. 274: 283–299.

The author expresses the view that the politicization of science is a relatively recent phenomenon that can be traced to excess of the administration of President George W. Bush. He examines some of the instances of this phenomenon that have been reported and suggests some basis for their occurrence.

Zucker, Kenneth J. 2003. "The Politics and Science of 'Reparative Therapy'." *Archives of Sexual Behavior.* 32(5): 399–402.

The author discusses the development and use of a technology supposedly capable of changing a person's sexual orientation, known as *reparative therapy*, for which no widely accepted empirical evidence ever existed, and how that procedure became accepted by a number of therapists and potential clients.

Reports

Ames, C. J., et al. *Report on the Misuse of Science in the Administrations of George H. W. Bush (1989–1993) and William J. Clinton (1993–2001).* http://cstpr.colorado.edu/admin/publication_files/resource-1935-2004.27.pdf. Accessed on July 31, 2013.

This report was compiled by members of the ENVS 4800 class at the University of Colorado in 2004. It is an attempt to provide a more specific understanding of the meaning of the "misuse of science" as described by two well-known studies on the issue, one by House of Representatives Committee on Government Reform and one by the Union of Concerned Scientists. The class members found that instances described as the "misuse of science" in these reports were of four types: mistakes, mischaracterizations, delegitimization, and arguing politics and/or morals through science.

Donaghy, Timothy, Francesca Grifo, and Meredith McCarthy. *Interference at the EPA: Science and Politics at the U.S. Environmental Protection Agency.* Cambridge, MA: Union of Concerned Scientists, 2008.

This report examines circumstances under which industry lobbyists and politicians have led to the suppression and distortion of research results from studies conducted by or under the auspices of the U.S. Environmental Protection Agency.

Grifo, Francesca, Michael Halpern, and Peter Hansel. *Heads They Win, Tails We Lose: How Corporations Corrupt Science at the Public's Expense.* Cambridge, MA: Union of Concerned Scientists, February 2012. Available online at http://www.ucsusa .org/assets/documents/scientific_integrity/how-corporations-corrupt-science.pdf. Accessed on May 10, 2013.

This report provides evidence of the ways in which corporations attempt to influence the use of scientific information at every step of the decision-making process in order to avoid regulatory actions by governmental agencies or to shape decisions in their favor.

Massarani, Tarek. *Redacting the Science of Climate Change: An Investigative and Synthesis Report.* Government Accountability Project. http://www.whistleblower.org/storage/documents/Redac tingtheScienceofClimateChange.pdf. Accessed on July 29, 2013.

This report summarizes a year-long study of the way in which governmental agencies in the United States have interfered with the collection and summary of data, the reporting of that data, and the use of the data for policy-making purposes in the area of climate change.

"Scientific Integrity: Let Science Do Its Job." http://www .ucsusa.org/scientific_integrity/. Accessed on August 2, 2013.

The Union of Concerned Scientists has conducted a number of studies to determine the extent to which scientists working in and for the federal government have had their work censured or blocked by their superiors. This web page provides access to those studies and to a number of other reports on the politicization of scientific research and reporting in the federal government.

Shulman, Seth. 2004. *Scientific Integrity in Policymaking: An Investigation into the Bush Administration's Misuse of Science.* Union of Concerned Scientists. http://www.ucsusa.org/assets/documents/scientific_integrity/scientific_integrity_in_policy_making_july_2004_1.pdf. Accessed on May 6, 2013.

This report reviews the evidence that members of the administration of President George W. Bush had consistently and extensively interfered with traditional procedures by which science-related decisions were being made at the highest levels of policy making.

"The Stem Cell Debates." 2012. *The New Atlantis.* Number 34: whole. Available online at http://www.thenewatlantis.com/publications/the-stem-cell-debates-lessons-for-science-and-politics. Accessed on August 3, 2013.

The Witherspoon Council on Ethics and the Integrity of Science sponsored a forum on federal policy on the use of human embryos in the development of stem cell research. This issue of the journal *The New Atlantis* reports on the science behind embryonic stem cell research, some suggested benefits of the research, ethical issues involved in the research, policy and law regarding stem cell research, and international aspects of the issue. The introduction to the issue notes the "strangeness of the debate" that saw "a government scientific advisory board strategizing about political lobbying, and politicians making passionate personal pleas about science policy."

United States House of Representatives. Committee on Government Reform—Minority Staff. Special Investigations Division. "Politics and Science in the Bush Administration." August 2003. http://oversight-archive.waxman.house.gov/documents/20080130103545.pdf. Accessed on July 31, 2013.

At the request of Rep. Henry Waxman (D-Calif.), the Committee on Government Reform reviewed instances in which the Bush administration had purportedly misused science for political motives. The committee found examples of such behavior in more than 20 agencies dealing with topics such as abstinence-only education, agricultural pollution, Arctic National Wildlife Refuge, breast cancer, condoms, drinking water, education policy, environmental health, global warming, HIV/AIDS, lead poisoning, missile defense, oil and gas, prescription drug advertising, reproductive health, stem cells, substance abuse, wetlands, workplace safety, and Yellowstone National Park.

Internet Resources

"2004 Scientist Statement on Restoring Scientific Integrity to Federal Policy Making." Union of Concerned Scientists. http://www.ucsusa.org/scientific_integrity/abuses_of_science/scientists-sign-on-statement.html. Accessed on May 6, 2013.

On February 18, 2004, a group of more than 60 leading scientists signed a petition taking note that the members of the administration of President George W. Bush had been unduly involved in misrepresenting, mischaracterizing, and even falsifying the results of scientific research used to make national policy decisions. The number of scientists signing this petition eventually reached more than 15,000.

["Allegation of scientific and scholarly misconduct and reprisal."] http://www.klamathbasincrisis.org/settlement/science/houser allegation022412.pdf. Accessed on May 6, 2014.

This document is a formal letter of complaint filed by Dr. Paul Houser against Secretary of the Interior Ken Salazar for alleged political interference in a scientific decision-making process about the removal of dams on the Klamath River, California. See Chapter 2 for more details about this issue.

Ball, Philip. "Science and Politics Cannot Be Unmixed." Homunculus. http://philipball.blogspot.com/2012/03/science-and-politics-cannot-be-unmixed.html. Accessed on August 3, 2013.

This London-based author takes on the president of the Royal Society for calling for the separation of science and politics, explaining in some detail that the mix between the two has existed at least for the last century or more.

Blank, Joshua. "Science and Politics—and Partisans." The Texas Tribune. http://www.texastribune.org/2012/11/05/science-and-politics-and-partisans/. Accessed on August 3, 2013.

The author, a PhD student at the University of Texas at the time of this article, reports on the results of a public opinion poll which shows a significant difference in attitudes between Republicans and Democrats about the trustworthiness of scientific data in the development of public policies, with the former largely regarding such data as untrustworthy and not very useful and the latter holding the opposite view.

Bolsen, Toby, James Druckman, and Fay Lomax Cook. "The Politicization of Science and Support for Scientific Innovations." http://www.ipr.northwestern.edu/publications/docs/workingpapers/2013/IPR-WP-13-11-REV.pdf. Accessed on July 26, 2013.

The authors present what they believe to be "the first *empirical* foray into how the politicization of science affects public opinion—in particular, support for the use of an innovative/still emergent technology [nuclear energy]." (Emphasis in the original.) They find that the politicization of science does, in fact, have negative effects on political opinion about at least the one form of scientific innovation that they studied.

Bradley, Robert Jr. "Draft National Assessment on Climate Change: Politicization of the Scientific Method." Master Resource. http://www.masterresource.org/2013/05/draft-national-assessment-pseudoscience/. Accessed on August 1, 2013.

The author of this web page argues that the authors of the National Assessment are picking and choosing among data, misrepresenting data, making improper assumptions and conclusions, and corrupting available scientific evidence about climate change in other ways, providing an example of the politicization of science by supposedly liberal government officials.

Breckler, Stephen. "Politics, Science and the Suppression of Research." American Psychological Association. http://www .apa.org/science/about/psa/2013/03/politics-suppression.aspx. Accessed on August 3, 2013.

The author writes about political pressures that prevent researchers from studying problems associated with gun violence and ways in which it can be reduced.

Briggle, Adam. "Let Politics, Not Science, Decide the Fate of Fracking." Slate. http://www.slate.com/blogs/future_tense/ 2013/03/12/fracking_bans_let_politics_not_science_decide .html. Accessed on August 3, 2013.

The author challenges the "conventional wisdom" in complex socioscientific issues like fracking to "let science dictate the final decision" on the issue. He points out that many such issues, including fracking, are largely a moral and social issue and that political forces, not science, should make the decision as to whether, when, and how fracking should be used for the extraction of fossil fuels.

Burkeman, Oliver. "Memo Exposes Bush's New Green Strategy." The Guardian. http://www.theguardian.com/environment/ 2003/mar/04/usnews.climatechange. Accessed on August 3, 2013.

This article discusses a 2003 memo written by Republican Party strategist Frank Luntz, in which he advises the Bush administration that the party has "lost the environmental communications battle," and it should switch discussion from known scientific information to a political approach. He points out in the memo that "A compelling story, even if factually inaccurate, can be more emotionally compelling than a dry recitation of the truth." (The memo itself can be found on http://nigguraths.files.wordpress. com/2013/03/luntzresearch_environment.pdf.)

Carter, Stacy M. "JAMA Forum: Separating the Science and Politics of 'Obesity'." News@JAMA. http://newsatjama.jama. com/2013/02/14/jama-forum-separating-the-science-and-pol itics-of-obesity/. Accessed on August 3, 2013.

> This article discusses the debate over a scientific article on the measurement of body weight in the diagnosis of obesity. The author points out that the debate is based not so much on the scientific aspect of the paper as on the variety of contexts in which the topic of obesity is discussed in today's world.

"Category Archives: Politicization of Science." Neutral Source. http://neutralsource.org/archives/category/topics/topics-regula tory_science/politicized-science. Accessed on July 30, 2013.

> This web site offers a variety of articles containing information and viewpoints on a new set of topics of current interest. This category includes articles on controlled substances, climate change and climate control, the Affordable Care Act of 2010, and the neutrality of scientists working on controversial topics.

Cohen, Steven. "The Science and Politics of Preserving Our Beach Communities." http://www.huffingtonpost.com/steven- cohen/the-science-and-politics_b_3305608.html. Accessed on August 1, 2013.

> The author focuses on one very specific example of the intersection between politics and science—the practice of building homes on unstable beach land—to discuss some general principles about the use of science in making political decisions, and the role of politicians in making use of scientific information, with some suggestions for improving the way the two fields interact with each other.

Dunning, Brian. "The Science and Politics of Global Warming." Skeptoid. http://skeptoid.com/episodes/4309. Accessed on August 2, 2013.

> This blogger takes on the task of trying to understand how the issue of global climate change became such a contentious issue among politicians and how the politicization of that issue has harmed the ability to make rational policy decisions and to conduct rational debates within the general public.

Hanson, Carla. "Politics, Science and Our (Not-So) Liberal Minds." BlueOregon. http://www.blueoregon.com/2013/04/politics-science-and-our-not-so-liberal-minds/. Accessed on August 3, 2013.

> For the fourth time in the last half century, the voters of the very liberal city of Portland, Oregon, rejected a ballot measure to begin fluoridation of the city's water supply. This blog examines the relative strength of scientific and political arguments in deciding how voters made up their mind about this issue. Perhaps the most interesting and useful part of the blog is the very extended commentary provided by many dozens of respondents to the original article.

Houser, Paul R. "Klamath Science-Informed Decision Making Process Needs Improvement." http://prhouser.com/houser/?p=830. Accessed on May 7, 2013.

> A hydrometeorologist working as a science advisor for the U.S. Bureau of Reclamation reviews his experiences in dealing with the way in which a governmental bureaucracy treated scientific data about a politically charged environmental issue, the removal of dams on the Klamath River in California.

"Interior Denies Spinning Klamath Science." Protecting Employees Who Protect Our Environment. http://www.peer.org/news/news-releases/2013/03/25/interior-denies-spinning-klamath-science/. Accessed on May 7, 2013.

This news story describes the actions taken in dealing with a complaint filed by Paul R. Houser, a science advisor who worked for the U.S. Bureau of Reclamation in assessing the environmental impact of removing dams on the Klamath River in California.

Johnson, Eric Michael. "The Failed Synthesis: Eduard Kolchinsky on the Dangers of Mixing Science and Politics." ScientificAmerican. http://blogs.scientificamerican.com/primatediaries/2012/06/12/failed-synthesis/. Accessed on August 3, 2013.

The author presents an interview with a man (Kochinsky) about the dangers that accrued in the Soviet Union as it attempted to interfere with the scientific process with such actions as the promotion of Lysenkoism in agricultural sciences.

Kloor, Keith. "The Poisoned Debate between [sic] Science, Politics and Religion." Discover. http://blogs.discovermagazine.com/collideascape/2012/12/27/the-poisoned-debates-between-science-politics-and-religion/#.Uf0-9ZLCZ8E. Accessed on August 3, 2013.

This article focuses on a number of issues that involve the interaction between science and politics, claiming that "Science and politics are intertwined, whether we like it or not." The writer's views on the topic elicit a very long list of responses and commentaries from others interested in the issue.

Kryder, Dylan Otto. "The Politicization of Science in the Bush Administration: Science-as-Public Relations." Skeptic. http://www.skeptic.com/eskeptic/04-10-08/. Accessed on July 26, 2013.

The author discusses examples of the ways in which members of the Bush administration have interfered with the normal process by which scientific information is obtained, distributed, and used for making policy decisions.

"Lamar Smith, GOP Push Politicization of Scientific Research." Huffington Post. http://www.huffingtonpost.com/ 2013/04/29/lamar-smith-science_n_3165754.html. Accessed on July 29, 2013.

> The author describes a bill submitted in the U.S. House of Representatives that would require the National Science Foundation to submit information on research proposals that it has approved to Congress for final review and action.

Long, Robert. "The Politics and the Science of Disputing Evolutionary Psychology." http://www.theamericanconservative .com/politics-and-science-of-disputing-evolutionary-psychol ogy/. Accessed on August 3, 2013.

> The author discusses the ways in which political issues have become involved in an ongoing dispute about a relatively new field of science known as evolutionary psychology.

Michaels, Patrick J. "The Politicization of Science Is Undermining the Credibility of Academia." The Cato Institute. http://www.cato.org/publications/commentary/politicization-science-is-undermining-credibility-academia. Accessed on July 26, 2013.

> The author argues that a "left-leaning bias of the social sciences" has produced a politicization of scientific research that is leading governments to the adoption of social policies that are not justified by the true nature of scientific research.

"Mixing Science, Politics Can Result in Bad Policy." The Oklahoman. http://newsok.com/mixing-science-politics-can-result-in-bad-policy/article/3732562. Accessed on August 3, 2013.

> This editorial provides a deniers view of global climate change issues, claiming that scientists are promoting a particular view of climate research for political reasons.

Mole, Beth Marie. "Q&A: Mixing Science and Politics: Politicians Could Make Better Decisions If They Thought More like Scientists, Says Rush Holt, the Only Physicist in Congress." The Scientist. http://www.the-scientist.com/?articles.view/articleNo/32775/title/Q-A--Mixing-Science-and-Politics/. Accessed on May 7, 2013.

Neuman, Scott. "War of the Worlds: When Science, Politics Collide." NPR News. http://www.npr.org/2012/04/13/150577766/war-of-the-worlds-when-science-politics-collide. Accessed on August 3, 2013.

> This report provides a brief history of the debate over the teaching of evolution in public schools with the most recent effort by antievolutionists in Tennessee to (once again) prevent the teaching of that subject in state schools.

"Obama Moves to Separate Politics and Science." CNN Politics. com. http://www.cnn.com/2009/POLITICS/03/09/obama .science/. Accessed on August 2, 2013.

> President Barack Obama was sensitive to the criticism of the administration of President George W. Bush from his earliest days in office. He decided to make sure that his administration could not be criticized for the same type of politicization of science by taking some of the steps described in this article.

O'Toole, Molly, and Munier Salem, "Science & Politics." Cornell Daily Sun. http://cornellsun.com/node/33101. Accessed on August 3, 2013.

> This article deals with the way in which political factors affected the teaching and practice of scientific research at an important university in a number of fields, including nuclear power and astronomy. The cases discussed are limited in application to Cornell, but have significance to any general discussion of the relationship between science and politics.

Pierce, William. "The Role of Science and Politics in Public Policy Decision Making." http://www.huffingtonpost.com/

william-pierce/the-role-of-science-and-p_b_1734920.html. Accessed on May 7, 2013.

> The author was spokesperson for the U.S. Department of Health and Human Services from 2001 to 2005. He argues in this essay that science itself is a political endeavor, with proponents on one or another side of controversies such as climate change or immunization policy. Therefore, it is legitimate for politicians themselves to take positions promoting one side or another in such debates.

"The Politicization of Science in Canada." http://individual. utoronto.ca/bernard/Politicization_of_Science_in_Canada/ Politicization_of_Science_in_Canada.html. Accessed on August 1, 2013.

> The politicization of science is hardly a unique American phenomenon; examples of this process can be found in many countries around the world. This web page provides 10 examples of recent events in Canada that illustrate the ways in which the scientific process can be and has been disrupted by political factors and influences.

"The Politics of Science." Center for Science & Technology Policy Research. http://sciencepolicy.colorado.edu/news/ media-resources/science-politics.html. Accessed on August 1, 2013.

> This web page provides a good general introduction to some of the fundamental ideas involved in a study of the politicization of science.

"Principles of Scientific Integrity." Coalition to Promote Research. http://www.cossa.org/CPR/cpr.html. Accessed on May 6, 2013.

> This web site is maintained by a group of scientific associations who are concerned about the potential for outside influences interrupting the normal process by which scientific research is proposed, approved, and funded in the United States.

"Project Apollo: A Retrospective Analysis." National Aeronautics and Space Administration. http://history.nasa.gov/Apollomon/ Apollo.html. Accessed on August 1, 2013.

> The NASA History Office has prepared this very detailed and very insightful review of the policy issues involved in the planning and development of the Apollo project.

"Recently in *Science and Politics* Category." Panda's Thumb. http://pandasthumb.org/archives/religion-and-po/science-and-pol/. Accessed on August 3, 2013.

> This web site contains a number of recent articles concerning the ongoing debate over the teaching of evolution, creation science, and intelligent design in American schools.

"Rescuing Science from Politics." Center for Progressive Reform. http://www.progressivereform.org/scienceRescue.cfm. Accessed on August 3, 2013.

> The center has long battled to protect the integrity of the scientific process and the work of scientists from inappropriate political intrusion. This article explains that position and provides links to more than a dozen specific efforts to carry out that objective in actual practice. An example is a book on the political influences acting on the formulation and conduct of public health policy written by center scholars.

Sabalow, Ryan. "Updated: Ousted Scientist Claims Salazar 'Spun' Klamath Dam Removal Reports." redding.com. http:// www.redding.com/news/2012/feb/28/whistleblower-says-science-ignored-siskiyou-county/. Accessed on May 6, 2013.

> This news report summarizes the controversy involving Paul R. Houser, who claims he was fired by Secretary of the Interior Ken Salazar for raising questions about the use of scientific data in the question of removing dams along the Klamath River in California.

"Science, I Choose You!" http://scienceichooseyou.wordpress.com/2013/03/13/science-vs-politics-in-canada-is-this-the-only-way/. Accessed on August 3, 2013.

> The controversy over political control of science has been at least as intense in Canada over the past decade as it has been in the United States. This blog provides an excellent review of some of the most important events that have occurred in that nation since the beginning of the 21st century, as well as public reaction to those events from around the world.

"The Science and Politics of Silica Sand Mining." PBS Videos. http://video.tpt.org/video/2337150233/. Accessed on August 3, 2013.

> This video deals with the scientific and political issues involved in the process of fracking. It includes interviews with the commissioner of the Minnesota Pollution Control Agency and a number of legislators attempting to develop policy on use of the procedure.

"Scientists, Politics, and Religion." Pew Research Center for the People & the Press. http://www.people-press.org/2009/07/09/section-4-scientists-politics-and-religion/. Accessed on August 2, 2013.

> This web page summarizes a number of poll results obtained by the Pew organization on the role of scientists in politics and the effects of political actions on scientific research.

Sheppard, Kate. "Perry Officials Censored Climate Change Report." http://www.motherjones.com/politics/2011/10/perry-officials-censored-climate-report. Accessed on May 9, 2013.

> A report on sea levels in Galveston Bay is edited by Texas state officials to remove mention of the possible role of global climate change in the situation under review.

Shermer, Michael. "Democracy's Laboratory: Are Science and Politics Interrelated?" Scientific American. http://www

.scientificamerican.com/article.cfm?id=democracys-laboratory. Accessed on August 1, 2013.

> The author reviews a proposition from Timothy Ferris' book, *The Science of Liberty*, which makes the argument that rather than democratic institutions providing the basis for scientific research, the situation is the reverse, with the scientific method serving as the progenitor of and providing the model for the operation of a democratic state.

Silver, Howard J. "Science and Politics: The Uneasy Relationship." Open Spaces Quarterly. 8(1). http://open-spaces.com/article-v8n1-silver.pdf. Accessed on May 5, 2013.

> The writer describes some recent instances in which political forces have interfered with the normal process by which government-supported research is approved and applied. He suggests that "charges of politicization—combined with a disregard for scientific evidence—has heightened the tensions between the scientific and political community."

Smith, Adam. "Science and Politics: Chalk and Cheese?" The Guardian. http://www.theguardian.com/science/2012/may/04/science-politics-chalk-cheese. Accessed on August 1, 2013.

> This article is the first in a series by the author describing the interrelationships of science and politics in the British governmental system. The articles are of special interest because of the similarities and contrasts between the British and American experiences.

Socolow, Robert, Roger A. Pielke Jr., and Randy Olson. "When Politicians Distort Science." Bulletin of the Atomic Scientists. http://www.princeton.edu/mae/people/faculty/socolow/bulletin-distortions-of-science-essays.pdf. Accessed on July 27, 2013.

> Three experts in the field of science exchange thoughts on the problems raised by the politicization of science, especially as illustrated by comments being made during the

young presidential campaign of 2011–2012, with special consideration as to how the treatment of the methods of science is likely to affect both science itself and the way that scientific information is or is not used by policy makers.

Tierney, John. "Politicizing Science." The New York Times. http://tierneylab.blogs.nytimes.com/2009/02/27/politicizing-science/?_r=0. Accessed on August 1, 2013.

The New York Times columnist and blogger Tierney raises the question as to what extent the politicization of science continues under the Obama administration, compared to the experiences of the Bush administration. Responses to Tierney's thesis are of particular interest in this blog.

Vergano, Dan. "Science vs. Politics Gets Down and Dirty." USA Today. http://usatoday30.usatoday.com/news/washington/2007-08-05-science-politics_N.htm. Accessed on May 7, 2013.

This column reviews the conflicts between science and politics during the presidential administration of President George W. Bush, focusing on stem cell research, global climate change, birth control, and endangered species. The column discusses in some detail the testimony of former surgeon general Richard Carmona before a committee of the U.S. House of Representatives in which he called the administration's policies "malicious, vindictive, and mean-spirited."

7   Chronology

The interaction between science and politics dates as far back in human history as any form of scientific investigation can be identified. Many events of significance have occurred throughout that long history, only a sample of which can be included in this chronology.

700 BCE   The rise of classic Greek science, and with it the first widespread use of basic research to learn more about the nature of the universe.

391 BCE   One possible date assigned to the burning of the great Library of Alexandria, motivated, according to some sources, by the hatred of a Christian mob of the wisdom contained in the library's books.

389 BCE   Possible date for Plato's *Republic*, in which he describes and defends the need for a "noble lie" which government leaders may use to teach citizens about their role in a stable society.

1543   The first edition of a book espousing the heliocentric theory of the universe, by Polish astronomer Nicholas Copernicus, is published just prior to his death. Historians believe that he held off publication of the book as long as possible to

A vending machine at Shippensburg University's Etter Health Center provides the Plan B emergency contraceptive along with condoms, decongestants, and pregnancy tests. (AP Photo/Shippensburg University)

avoid condemnation and punishment by the Roman Catholic Church because of its heretical ideas.

1600   Italian philosopher and natural scientist Giordano Bruno is burned at the stake in Rome for holding views about the nature of the universe contrary to those approved by the Roman Catholic Church.

1603   Founding of the world's first scientific society, the Accademei dei Lincei.

1620   Sir Francis Bacon publishes his description of the rules for the conduct of science and the purposes for which scientific research ought to be conducted in *The Great Insaturation.* His even more famous work on the subject, *Novum Organum*, is published in the same year.

1623   Galileo lays out many of the principles of modern science in his book *Il Saggiatore* (*The Assayer*).

1633   Galileo is tried before inquisitor Cardinal Vincenzo Maculani for holding views about the world contrary to the teachings of the Roman Catholic Church. He is sentenced to imprisonment, which is almost immediately commuted to house arrest for life.

1637   French physicist René Descartes publishes *Discours de la méthode* (*Discourse on the Method*) describing his recommendations for the procedures by which scientific research should be conducted.

1660   Establishment of the Royal Society of London, the world's oldest continuous scientific society.

1665   Publication of the first continuous scientific journal in the world, *Journal des sçavans*, first published by the Académie des Sciences in January 1665, followed two months later by the first issue of the English *Philosophical Transactions of the Royal Society.*

1666   Establishment of the French Académie des Sciences, the world's second oldest continuous scientific society.

1712   English inventor Thomas Newcomen produces the first successful model of the steam engine that was to mark the beginning of the Industrial Revolution.

1787   The French Revolution begins, a period in which science takes a central role in the transformation of French society.

1802   The U.S. government creates the Corps of Engineers, the first federally sponsored scientific agency. (Some authorities assign that honor to the Survey of the Coast, established five years later.)

1807   The Survey of the Coast, forerunner of the modern National Oceanic and Atmospheric Administration (NOAA), is created, credited by many authorities as the first federal scientific agency.

1830   In his book, *Reflections on the Decline of Science in England*, English mathematician Charles Babbage outlines a number of reasons for the decline of scientific research in the country, blaming government's ignoring research as one important factor.

1846   The Smithsonian Institution is established in Washington, D.C., with a grant from English scientist James Smithson. The institution is the first U.S. agency devoted primarily or exclusively to the support of basic research in science.

1876   The Johns Hopkins University, the nation's first primarily research institution of higher learning, is established in Baltimore, Maryland.

1902   American philanthropist and businessman Andrew Carnegie endows the Carnegie Institution, whose primary purpose was the support of basic research in the sciences.

1925   For the first time, drug experts list marijuana with cocaine, heroin, morphine, opium, and coca as a high-risk substance for human use.

1925   The Tennessee legislature passes the Butler Act, banning the teaching of human evolution in public schools in the state. That law is tested in a July court case, *State of Tennessee v. John Thomas Scopes*, in which high school biology teacher John Scopes is found innocent on a technicality.

1929–1931   The height of the purge of scientists with unacceptable political views occurs in the Soviet Union.

1931   Representatives from the Soviet Union at the International Congress of the History of Science and Technology in London present a series of papers, later collected in the form of the book *Science at the Crossroads*, showing how government management of science has been successful in solving a host of social problems in the country.

1932   The National Socialist (Nazi) party in Germany wins a majority in national elections, allowing it to accelerate its programs against Jews in the country and initiating its efforts to remove Jews from all teaching, research, and other scholarly positions in the country.

1933   The German parliament adopts the Law for the Reestablishment of the Professional Civil Service, removing Jews from government service. It also passes the Law against Overcrowding in Schools and Universities, limiting the number of Jewish students permitted in schools and universities.

1936   Trofim Lysenko is appointed director of the All-Union Institute of Selection and Genetics and chosen to be president of the All-Union Academy of Agricultural Sciences, giving him the authority required to introduce his (faulty) theory of genetics, proposed to solve the Soviet Union's desperate food shortages at the time.

1936–37   German physicist and Nobel Prize winner Philipp Lenard publishes his four-volume textbook on physics, *Deutsche Physik* (*German Physics*), based on the concept that the facts of science may differ depending on the ethnic background of individuals by whom they are produced.

1939   Crystallographer and social activities J. D. Bernal publishes *The Social Function of Science* explaining why and how science must be brought under government control in order to solve a nation's social problems.

1940   British cytologist John Baker and Hungarian-born philosopher of science Michael Polanyi found The Society for Freedom in Science, an organization designed to promote the concept of and conditions that make possible free and unhindered basic research by scientists.

1942   President Franklin D. Roosevelt authorizes the initiation of a project to determine the feasibility of building nuclear weapons, a project to become known as the *Manhattan Project.*

1944   A report produced by the New York Academy of Medicine on the medical and health effects of smoking marijuana finds no significant medical or health effects from use of the drug by humans.

1945   A group of scientists who had been working on the Manhattan Project organize to form the Federation of Atomic Scientists, which soon changes its name to the Federation of American Scientists.

1953   American physicist J. Robert Oppenheimer, intellectual father of the Manhattan Project to build the first nuclear weapons, is stripped of his security clearance. The action is taken supposedly because of his sympathy toward Communist causes but, in fact, is the result of Oppenheimer's objections to existing governmental policies about the development and use of nuclear weapons.

1961   President John F. Kennedy announces that the United States will attempt to put a man on the Moon before the end of the decade. Historians have pointed out that the President had little or no interest in the scientific component of this project, but that he conceived the project as a way of showing superiority over the Soviet Union in the area of space research.

1968   The U.S. Supreme Court rules unanimously in the case of *Epperson v. Arkansas* that state laws prohibiting the teaching of human evolution are unconstitutional.

1969   Astronauts Neil Armstrong and Buzz Aldrin land on the surface of the Moon, bringing a successful conclusion to President Kennedy's promise eight years earlier.

1969   A group of faculty and students at the Massachusetts Institute of Technology (MIT) organize to form the Union of

Concerned Scientists (UCS) to, among other objectives, inform policy makers and the general public about the interaction of science and politics and specific issues that may arise out of that interaction.

1970   The U.S. Congress passes the Controlled Substances Act, which categorizes marijuana as a Schedule I drug, having no medical benefits and high risk for harm to humans who use it.

1972   The National Commission on Marihuana and Drug Use (the "Shafer Committee") finds no significant health or medical effects from smoking of the drug and recommends decriminalization for possession of small amounts of the drug.

1975   Sen. William Proxmire (D-Wis.) announces the first of his "Golden Fleece" awards, grants made by the federal government to scientists for research that he considers to be frivolous and a waste of taxpayer money.

1981   The U.S. Congress passes the Adolescent Family Life Act (AFLA) emphasizing chastity and self-discipline as the primary means of avoiding pregnancy. The act is the first in a long series of actions by the Congress on behalf of abstinence-only education (AOE) programs in the United States.

1981   The State of Louisiana passes the Balanced Treatment for Creation-Science and Evolution-Science in Public School Instruction Act, requiring all public schools in the state to offer instruction in so-called creation science along with any teaching of organic evolution.

1981   A group of private citizens concerned about the spread of creationism in public schools found the National Center for Science Education, which is formally incorporated two years later.

1983   President Ronald Reagan announces the creation of the Strategic Defense Initiative (SDI), for the purpose of developing a "shield" of missiles to protect the United States from attack by foreign missiles.

1987   Reagan approves plans to build the Superconducting Super Collider (SSC), at an estimated cost of $4.375 billion.

1987 The U.S. Supreme Court rules in the case of *Edwards v. Aguillard* that the Louisiana law requiring the teaching of creation science (1981) is unconstitutional. The ruling is a 7 to 2 decision, with Chief Justice William Rehnquist and Justice Antonin Scalia dissenting.

1989 The U.S. Congress passes the Whistleblower Protection Act, whose primary goal is to protect federal employees for retribution by the agencies for whom they work because of information they provide about improper activities within those agencies. But also see *Garcetti v. Ceballos* (2006).

1989 President George H. W. Bush decides to abandon the SDI project in favor of a more limited missile defense system called Brilliant Pebbles.

1989 Opponents of the teaching of evolution in schools propose the theory of intelligent design, which they suggest should be taught in concert with evolution in public schools.

1991 Construction on the SSC begins at a site near Waxahachie, Texas.

1993 The U.S. Congress cancels the SSC project after 22.5 kilometers (14 miles) of the tunnel has been built and about $2 billion has been spent on the project.

1996 The U.S. Congress passes the Temporary Assistance for Needy Families Act (TANFA), establishing new grants for AOE and creating a detailed definition as to the meaning of AOE, now known as the A-H definition.

1996 A group of federal employees organize to form Public Employees for Environmental Responsibility, an organization whose primary purpose is to protect federal employees from retribution when they speak out against political interference in the process of scientific research.

1999 The Food and Drug Administration (FDA) approves the prescription use of the drug levonorgestrel as a contraceptive.

2003 The Special Investigations Division of the U.S. House of Representatives Committee on Government Reform issues a report that claims that members of the administration of President George W. Bush had interfered with scientific activities in at least 21 areas of government, such as agricultural pollution, Arctic National Wildlife Refuge, breast cancer, condoms, international negotiations, drinking water, education policy, and environmental health.

2004 A report by the U.S. House Committee on Oversight and Government Reform finds that two-thirds of AOE programs funded by the U.S. government provide incorrect information about a number of topics, including sexually transmitted infections, condom failure, and use of contraceptives.

2004 An advisory committee to the FDA votes 23 to 4 to endorse the use of the Plan B contraceptive drug for women of all ages without prescription.

2004 The Union of Concerned Scientists (UCS) releases a report on the treatment of scientific issues by the Bush administration entitled *Scientific Integrity in Policymaking: An Investigation into the Bush Administration's Misuse of Science.* The findings of the report are later published in book form as *The Republican War on Science* (Basic Books, 2005)

2004 The UCS initiates its Scientific Integrity program, with the goal of uncovering specific examples of political interference in scientific research by the federal government. The program has produced a significant amount of data about the type and number of instances of such interference that occur in a wide variety of federal agencies.

2005 In the case of *Kitzmiller v. Dover Area School District,* the U.S. District Court for the Middle District of Pennsylvania rules that the teaching of intelligent design in public schools (see 1989) is unconstitutional.

2005 The FDA decides to ignore the advice of its advisory committee (see 2004) and decides to make Plan B available without prescription to women over the age of 16 only. The agency later raises this age to 18.

2005   Senior associate in the U.S. Climate Change Science Program, Rick Piltz, resigns in protest over political interference by supervisor Philip Cooney and others in reports produced by the program. Shortly thereafter, Cooney himself also resigns from his post as chair of the Council on Environmental Quality (CEQ).

2006   In the case of *Garcetti v. Ceballos*, the U.S. Supreme Court rules that federal employees are not protected from retribution by their supervisors for whistle-blowing acts because they are acting within the provisions of their job descriptions. The decision essentially eviscerates the 1989 Whistleblower Protection Act (*q.v.*).

2007   Deputy Assistant Secretary for the Fish and Wildlife Service (Julie McDonald) resigns following accusations that she systematically interfered with research on and the writing of reports about scientific topics within her jurisdiction, although she had no special expertise in those areas.

2007   A group of concerned scientists organizes the Coalition on the Public Understanding of Science (COPUS).

2008   The Louisiana legislature passes the Louisiana Science Education Act, which allows public school teachers to use supplementary materials in science classes that are critical of established scientific conclusions about topics such as climate change and evolution. As of early 2014, the law has survived attempts at repeal within the legislature.

2009   Under orders by federal judge Edward Korman, the FDA lowers the age at which women can obtain Plan B without prescription to 17.

2009   Officials in the administration of President George W. Bush rewrite a report on the state of the environment, removing suggestions that climate change is likely to have significant consequences of human life on Earth and, instead, emphasizing a lack of consensus among scientists about the nature of global climate change, its causes, and its potential effects.

2009   Rep. Paul Broun (R-Ga.) says on the floor of the House of Representatives that "the idea of human-induced global climate change is one of the greatest hoaxes perpetrated out of the scientific community. It is a hoax. There is no scientific consensus." At the time, there is a general consensus among climate scientists worldwide that global warming is now an established fact.

2011   The FDA decides to make Plan B available to women of all ages without prescription. On the same day the decision is made, Secretary of Health and Human Services Kathleen Sebelius overturns the FDA ruling, and keeps the drug on prescription status for women over the age of 16.

2011   A report written by the Houston Advanced Research Center on sea level changes in Galveston Bay, prepared for the Texas Commission on Environmental Quality, is edited by the commission to remove any mention of the possible effects of global climate change on sea levels in the bay.

2011   President Barack Obama attempts to cut funding for AOE to $5 million in his 2012 budget, but the U.S. House of Representatives increases that amount to $50 million.

2012   U.S. Representative Todd Akin (R-Mo.), candidate for the U.S. Senate, claims that a woman's body naturally rejects the opportunity to become pregnant in cases of "legitimate rape."

2013   Judge Korman orders the FDA to make Plan B available to women of all ages without prescription within 30 days. The agency does so the following day.

Understanding the terms used in talking about the interrelationship between science and politics helps to understand the issues arising out of that relationship. Some of the most important of those terms are defined here.

abstinence   Voluntary avoidance of some activity, such as sexual activity.

abstinence-only education   A form of sex education in which abstinence is taught as the only certain method for preventing pregnancy outside of marriage and the only certain way of avoiding sexually transmitted infections. For a formal detailed definition of the term, see http://www.ssa.gov/OP_Home/ssact/title05/0510.htm.

applied science   Research that is carried out for the purpose of solving some specific practical problem in everyday life.

Aryan science   *See* **German science**

basic science   Research that is carried out for the purpose of gaining new knowledge about some natural object or phenomenon, without concern as to what, if any, practical application that knowledge may ultimately have.

A warning sign is shown attached to a fence at the "C" Tank Farm at the Hanford Nuclear Reservation, near Richland, Washington. Time is running out on Hanford's deteriorating tanks and for the federal government to complete work on a more permanent solution to store the radioactive materials within them. (AP Photo/Ted S. Warren)

big science A form of scientific activity that involves large numbers of scientists working on a research project with a very high cost, generally involving the use of large and expensive equipment. *Also see* **little science**.

cherry picking The act of choosing from a large amount of data only those points that support a person's individual views on a topic.

creation science A purported type of scientific endeavor in which scientific facts and theories are completely coordinated with Biblical explanations of the origin and development of the universe and everything within it. U.S. courts have determined that creation science is not a legitimate form of science. *Also see* **creationism; intelligent design**.

creationism The theological teaching that the universe and everything within it were created by the act of some supreme being. *Also see* **creation science; intelligent design**.

deligitimization The attempt to call into question the validity of the remarks made by an individual or an organization, or to call into question the validity of the person or organization itself.

dueling experts A phenomenon that occurs in court cases or in public debates over an issue when two or more putative experts present significantly different interpretations of data presented in a case.

gateway theory of drug use A theory that the use or abuse of one particular drug (such as marijuana) predisposes the user to an increased risk for the use or abuse of other, usually more dangerous, drugs (such as heroin or cocaine).

German science A belief promulgated by certain German scientists and members of the Nazi party in the 1930s that the laws of science are dependent not only on research and observation, but also on a researcher's individual racial, ethnic, and other characteristics. The theory was set forward as a contrast to Jewish science, which was described as at least partially false because it had been developed under the influence of individuals with untrustworthy genetic backgrounds (i.e., Jews). Also known as **Aryan science** and **Nordic science**.

intelligent design   A theory that the universe as it appears today, and all objects within it, were created at some time in the past by some intelligent force, such as a supreme being with the characteristics of a god or supernatural force.

Ionian science   A name sometimes given to a field of intellectual endeavor that arose in Greece in about the sixth century characterized by certain formal methods for collecting information about the natural world.

Jewish science   *See* **German science**

legitimate rape   A term that has sometimes been used in the debate over contraception and abortion with an unclear meaning. It seems most often to refer to women who have, in fact, been raped, in comparison to women who claim to have been raped (but were not actually raped) as a way of explaining one's pregnancy and is, therefore, "illegitimate rape."

little science   A term sometimes used to describe older, traditional forms of scientific research that can be carried out by a solitary scientist or a small group of scientists at relatively modest cost, usually with relatively simple equipment.

Luddite   Originally, someone in 19th century Great Britain who opposed technological changes taking place at the time because he or she feared such changes would reduce employment. The Luddite movement was named after one of the first such individuals, Ned Ludd, supposedly a weaver from Leicestershire, who carried out acts of vandalism against mechanized knitting frames. The term is now used for any individual who is opposed to technological change in general.

Lysenkoism   A biological theory originally proposed by Soviet agriculture scientist Trofim Lysenko based on the (incorrect) assumption that traits developed by an organism during its life can be transmitted genetically to its offspring.

mischaracterization   The intentional or unintentional explanation of a body of knowledge.

missile gap   A term used to describe the purported disparity between the number of long-range missiles held by the Soviet Union compared to the United States or, more generally, to the supposed military superiority of the Soviet Union over the United States in the early 1960s.

noble lie   An untrue story told by someone in authority about an important social concept that is important enough for the stability of a society to be expressed, even if false. Plato is often credited for having first described such statements.

Nordic science   *See* **German science**

politicization   The process of applying the methods usually associated with politics to other fields in which they may or may not be appropriate, such as religion or science.

politics   The theory and practice of influencing citizens' beliefs and thoughts about civic issues and of translating those beliefs and thoughts into formal laws and policies at all levels of governance.

pure science   *See* **basic science**

quality science   *See* **sound science**

schedule   In the terminology of substance abuse, a category of drugs with specific physiological and medical characteristics. The Controlled Substances Act of 1970 established five schedules of drugs based on their potential for misuse, possible medical benefits, and safety when used under proper supervision.

science   A way of describing the natural world that makes use of certain specific ways of collecting, interpreting, and distributing information collected by means of controlled observation and experimentation.

sound science   A term for which there is no universally accepted definition. The term is often adopted by individuals and groups with specific political biases to describe the type of scientific research of which they most highly approve, whether or not it corresponds to the traditional format of the scientific method. Also sometimes described as *quality science*.

Star Wars Among other meanings, the term was applied as a nickname for the Strategic Defense Initiative (SDI) missile defense system proposed by President Ronald Reagan in 1983.

technology The use of known scientific information for the design and production of objects useful in everyday life.

About the Author

David E. Newton holds an associate's degree in science from Grand Rapids (Michigan) Junior College, a BA in chemistry (with high distinction) and an MA in education from the University of Michigan, and an EdD in science education from Harvard University. He is the author of more than 400 textbooks, encyclopedias, resource books, research manuals, laboratory manuals, trade books, and other educational materials. He taught mathematics, chemistry, and physical science in Grand Rapids, Michigan, for 13 years; was professor of chemistry and physics at Salem State College in Massachusetts for 15 years; and was adjunct professor in the College of Professional Studies at the University of San Francisco for 10 years. Previous books for ABC-CLIO include *Global Warming* (1993), *Gay and Lesbian Rights—A Resource Handbook* (1994, 2009), *The Ozone Dilemma* (1995), *Violence and the Mass Media* (1996), *Environmental Justice* (1996, 2009), *Encyclopedia of Cryptology* (1997), *Social Issues in Science and Technology: An Encyclopedia* (1999), *DNA Technology* (2009), and *Sexual Health* (2010). Other recent books include *Physics: Oryx Frontiers of Science Series* (2000), *Sick!* (4 volumes) (2000), *Science, Technology, and Society: The Impact of Science in the 19th Century* (2 volumes; 2001), *Encyclopedia of Fire* (2002), *Molecular Nanotechnology: Oryx Frontiers of Science Series* (2002), *Encyclopedia of Water* (2003), *Encyclopedia of Air* (2004), *The New Chemistry* (6 volumes; 2007), *Nuclear Power* (2005), *Stem Cell Research* (2006), *Latinos in the Sciences, Math, and Professions* (2007), and *DNA Evidence and Forensic Science* (2008). He has also been an updating and consulting editor on a number of books and reference works, including *Chemical Compounds* (2005), *Chemical Elements* (2006), *Encyclopedia of Endangered Species* (2006), *World of Mathematics* (2006), *World of Chemistry* (2006), *World of Health* (2006), *UXL Encyclopedia of Science* (2007), *Alternative Medicine* (2008), *Grzimek's Animal Life Encyclopedia* (2009), *Community Health* (2009), and *Genetic Medicine* (2009).